Albert John Thomas Morris

A Treatise on Meteorology

The barometer, thermometer, hygrometer, rain-gauge, and ozonometer; with rules and regulations to be observed for their correct use

Albert John Thomas Morris

A Treatise on Meteorology
The barometer, thermometer, hygrometer, rain-gauge, and ozonometer; with rules and regulations to be observed for their correct use

ISBN/EAN: 9783337378080

Printed in Europe, USA, Canada, Australia, Japan

Cover: Foto ©berggeist007 / pixelio.de

More available books at www.hansebooks.com

A

TREATISE ON METEOROLOGY:

THE

BAROMETER, THERMOMETER, HYGROMETER, RAIN-GAUGE, AND OZONOMETER;

WITH

RULES AND REGULATIONS TO BE OBSERVED
FOR THEIR CORRECT USE.

TO WHICH ARE APPENDED SOME OF THE LATEST DISCOVERIES AND THEORIES
OF SCIENTIFIC MEN RESPECTING VARIOUS SOLAR AND
TERRESTRIAL PHENOMENA.

DEDICATED TO THE METEOROLOGISTS OF STRATHEARN.

BY

ALBERT J. T. MORRIS,
(QUEEN'S COLLEGE, OXFORD),
INCUMBENT OF ST JAMES' CHURCH, MUTHILL, PERTHSHIRE.

EDINBURGH:
R. GRANT AND SON, PRINCES STREET.
MDCCCLXVI.

PREFACE.

THE following Meteorological data are mainly derived from such well-known sources as Messrs SCOFFERN and LOWE on Practical Meteorology; Mr GLAISHER's Hygrometrical Tables; Sir W. ARMSTRONG's Presidential Address to the British Association, 1863; together with certain discoveries and *un*certain theories, propounded by Professor Tyndall, Mr Nasmyth, Sir John Leslie, and others; all of which we would gladly accept with implicit faith, if only their authors would always agree not to differ.

A very valuable treatise, on this same subject, was published a few years ago by Messrs Scoffern and Lowe; but, in many respects, it is far deeper than the ordinary meteorologist will require, embracing optics, acoustics, electricity, and other scientifically treated matters, for the full understanding of which the reader should have dipped into " the Mathematics," more deeply than the amateur meteorologist will generally be found to have done.

I therefore presume to offer this more familiar style of composition for the convenience of those who may wish for an easy Hand-book to the right employment of the few meteorological instruments commonly in use.

It will probably be thought that the facts supplied are of the greater value, in that I do not pretend to offer them as the original fruits of my own mind, but merely attempt to exhibit them in such a form as may commend itself by its brief simplicity.

A. J. T. Morris.

Balwharrie, near Crieff,
 30th March 1865.

After recording the long sad list of ships and human lives lost in the terrible storm which reached its height on the 10th of January 1866, when the noble steam-ship "London," with 220 passengers, foundered in the open sea, the "Edinburgh Evening Courant" makes the following appeal:—

"We trust that this sad narrative will not be thrown away on the shipping enterprise of this country; or, above all, on the votaries of meteorological science. If that science were in the state of advancement in which it ought to be, and to which it is certainly destined to attain, the hurricane which sweeps the ocean, and overwhelms our most seaworthy ships, could be, if not averted, at least foreseen and shunned.

"There is no science which is so abundant in facilities for its prosecution; none which enlists in its service so manifold and so various a body of auxiliaries. Were all the ships which cover our seas and coasts to be adequately supplied with meteorological observers,—were landsmen in every sea-port town, and in every convenient inland situation, to teach themselves how to make and to register meteorological observations, an induction of facts would soon be accumulated, which would enable us, not only to predict the coming of a storm, but even the force with which it would be charged.

" Governments have done much for the advancement of the sciences on which the security of our maritime enterprise depends; but governments may easily do more. They have contributed to protect the ship of the mariner from the lighting bolt,—they have planted beacons and light-houses on the headlands, to ensure his safe entrance into harbour,— and they have established the life-boat apparatus at every point menaced by the storm.

" Is it too much to ask that an equal impetus be given to that meteorological science, which is destined, in a great measure, to supersede the necessity of relying on such uncertain aids in the hour of extreme peril? Through the voice of the departing storm, let humanity and wisdom reply."

It may very well be, that the "Evening Courant" will fail to rouse government into anything like action in this matter, but its appeal to the humanity of our countrymen will not be thrown away, even though to respond to that appeal, by devoting themselves to the science of the winds, will demand of them a long patience. For the meteorologist cannot measure his progress in science by the results of weeks or months, but is content to know, that the carefully recorded phenomena of five or ten or fifteen successive years, may help to point out to some more gifted genius than himself the certain laws which govern the apparently uncertain elements. It is with such a hope that I venture to commend this little work to the favourable notice of both friends and strangers.

Happily, we are not, one and all, so slavishly bound down to the never-ending task-work of an iron age, but that we can call some of Time's hours our own, and may devote them to a study which, whilst it aims at the public good, also reveals new and unsuspected charms to the student, in proportion as the passing years remind him that he himself is growing old.

I cannot close these introductory remarks without thanking Stewart Hepburn, Esq., of Colquhalzie, for various suggestions and explanatory notes, and for the hearty way in which, after perusing the MS., he signified his approval of my small labours.

A. J. T. M.

January 22, 1866.

TREATISE ON METEOROLOGY.

ATMOSPHERE.

The atmosphere is the subject with which the meteorologist has to deal, with all the ever-changing phenomena of pressure, temperature, drought, and moisture, which it is constantly bringing before his notice.

In the widest sense of the word, the atmosphere is that mass of thin, elastic fluid by which any body is entirely surrounded; as when we speak of the atmosphere of the sun, moon, or planets, the existence of which may not be absolutely certain, though it is highly probable. Our earth may be said to swim through space, in company with an all-surrounding atmosphere of air and watery vapour, inseparably attached to the revolving body by the attractive principle which, as it were, draws all outer things together by some invisible central force. When speaking of this general tendency of all bodies to draw towards each other, with greater or less force, according to their density, we commonly merge cause and effect in one, by the use of the terms "heavy" and "weight."

By means of its weight, then, the air is inseparably connected with the earth, and presses on it in accordance with the laws of heavy elastic fluids,—its pressure being equally exerted on all sides, and in every possible direction. If now, from any cause, a stronger pressure is exerted in one direction, certain resulting phenomena arise in some other direction, and continue to act until the equilibrium is restored. Thus, for instance, water ascends in the bore of a pump, above its general level, as soon as a vacuum is created between it and the ascending piston. The cause of this is the disturbance of the equilibrium, since the air outside the bore presses on the water outside, whilst no air is present within the bore to counterbalance the outer pressure. By means of this pressure—if the tube be long enough—the water may be raised to the height of thirty-two and a-half feet.

This column of water in the tube represents the weight with which the atmosphere presses on the earth; or we may speak of the whole atmospheric pressure as being equal to an ocean of water thirty-two and a-half feet deep, spread over the entire surface of the earth. Hence it is calculated that with the barometric column standing at about twenty-eight inches, the atmosphere presses with a weight of 32,400 lbs. on the human body, reckoning the body's surface at fifteen square feet. The man, however, is unconscious of this enormous pressure, because the air surrounds him, and is, besides, *within* him. On account of its elasticity, it presses in every direction, even from within a man outwards, and in this way counterbalances the air spread externally over the body.

The atmosphere is not everywhere of uniform density; for the lower strata of the air, having to support the weight of the upper ones, become more compressed and dense. According to the law of Mariotte, the density of the atmosphere diminishes in geometrical, as the height increases in arithmetical, progression; or we may state the law thus, viz., that "the compression of the air is inversely to the pressure applied,"—that is to say, if the pressure be doubled, the volume of the air is reduced to one-half; if tripled, it is reduced to one-third; if quadrupled, to one-fourth, and so on. This law will not, probably, hold true at the extreme limits of our upper air, because, at that height the air being more free from superincumbent pressure must be more completely in its natural state.

The height of the atmosphere has been variously estimated at from thirty to fifty miles, calculating partly from the pressure which it exerts, and partly from the duration of twilight. For it may be supposed that the very loftiest stratum of air which reflects light upon us, before sunrise or after sunset, belongs to our planet. It is not of equal height, however, over all parts of the earth, but may be regarded as a spheroid, being most elevated at the equator, both on account of the centrifugal effects of our earth's diurnal motion, and on account of the expansive action of the sun's vertical rays, which is at its maximum between the tropics.

Nor is the atmosphere by any means the simple element which the ancients believed it to be, but is made up of several gases, invariably found in the same proportions everywhere, together with foreign ingredients unequally distributed, and floating about until again deposited on the earth, from whence they at first arose.

Mention will be made of these constituent and accidental parts of the air later on. Enough has now been said to prepare the way for the consideration of our first instrument, the barometer.

BAROMETER.

The barometer is the instrument by which changes in the weight of the air are ascertained.

The pressure of the superincumbent air, at one and the

same place, would always be alike, or nearly so, were it not that lateral disturbances remove a portion of the pressure at one point and add to it elsewhere. Thus, supposing the average pressure to be 29·70 in inches of a mercurial column,—then, if the barometer rises to 30·20 inches, or upwards, it is because a corresponding *fall* has taken place in some other part of the earth.

Hence it is that a very sudden rise of the barometric column is nearly always followed by a speedy fall, and *vice versâ*,—backwards and forwards,—each time less and less, until, by these oft-repeated undulations, the equilibrium of the whole air, over an area of some thousands of square miles, is at last attained, when a steady barometer and calm weather, more or less continuous, are the result.

Effect of Elevation on the Barometric Column.—The greater the perpendicular height above sea-level, the less will be the pressure of the air, simply, or rather chiefly, because there is less air above to exert this downward pressure.

By the application of this principle,—with corrections and reductions for temperature and elevation above sea-level,—we are able to judge, pretty correctly, as to the height of a mountain; and if mines could be sunk thousands instead of hundreds of feet, below the surface of the earth, the same principle, applied the contrary way, would give us a tolerably accurate idea of their depth. Indeed, if all the necessary precautions and allowances are observed, and the instrument be a good one, the height of a mountain may be learnt, within five or six feet, by barometric measurement, as certainly as by a trigonometric survey.

At an elevation of about thirty-six miles, the pressure of the atmosphere is probably not more than 0·001, or the one-thousandth part of an inch of the barometric column; in other words, the mercury, at such an elevation, would sink down and leave the barometer-tube empty, because the air would be of such a rarity as to be nearly a vacuum exerting no pressure. Conversely, at the depth of about sixty-six miles below the earth's surface (supposing a mine sunk to that depth, and left open for the ingress of the air !) the density of the air would be about 100,000 times greater than at the sea-level, or six times more than the density of gold,—so that a lump of gold would actually float very buoyantly in such an atmosphere.

Elevation of Observers' Stations.—On the sea-coast it is probable that an observer's house may not be many feet (say thirty or forty feet) above sea-level;—more inland he may be, as at Crieff, 250 feet above the sea; whilst at the Glasgow Waterworks' station, between Glenfinlas and Ben Ledi, he may be as much as 1800 feet above the sea.

Supposing, then, that three good standard barometers, all alike as to their "reading" at sea-level, were placed at these different elevations, their mercurial columns would stand at very different points :—

The one at the sea-coast might stand, say, at 30·00 inches.
That at the elevation of 250 feet would
be at . . . . . . 29·75 „
That at the elevation of 1800 feet would
be at . . . . . . 28·20 „

And yet, at all these places, the weather, as we popularly term it, might be equally good and steady; and the only reason for the very low barometer at Glenfinlas is, that its elevation is 1800 feet above the sea.

Rules for Allowances.—Now, it is obvious that if observers, so differently situated, are to take observations, capable of being compared with each other, and with such head-quarters as Greenwich, Liverpool, Edinburgh, and Glasgow, they must, one and all, adopt the same rules of allowances.

Two such rules are now recognised, as sufficiently accurate for all the practical objects of the meteorologist, viz.—

1st, *The allowance for elevation above sea-level.*
2d, *The allowance for temperature.*

In the meteorological statement which Admiral Fitzroy first published daily in the "Times" (and which is still reported there by the authority of his successor), he thus expresses it :—

"*Barometer Corrected and Reduced.*

"Each ten feet of vertical elevation, above half-tide level, causes about one-hundredth of an inch diminution; and each 10° above 32° causes nearly three-hundredths of an inch increase." (See any copy of the "Times.")

If we wished to find the height of a lofty mountain (say, 15,000 feet high) the above rules of allowance would *not* be sufficiently precise for our purpose, but would leave us, perhaps, some hundreds of feet wide of the mark. But they are accurate enough for the stations of observers in this country, which are rarely 1000 feet above sea-level; and being now recognised as the rules for allowance, it is clear that any little inaccuracy is at least counterbalanced by the advantages of uniformity.

Of course, 32° (Fahr.) as the starting-point is partly arbitrary; any other temperature might have been chosen as the standard. A rule is often made more clear by some familiar example. We will therefore suppose a case which shall take in both allowances requisite, viz., for elevation and for temperature.

Required to know the true altitude of the mercury at the level of the sea, and at the temperature of 32° (Fahr.) :—

Let the observation be made at Crieff, 250 feet above the sea;
Let the temperature of the barometer-room be 55° (Fahr.);

BAROMETER.

Let the *apparent* height of the mercurial column be 29·565 inches;

Then the *apparent* reading is . . . 29·565

250 × 0·001 = 0·250, and gives . $\{\substack{\text{plus}\\+}\}$ 0·250 $\{$ (For elevation.)

—————

29·815

But 55° − 32° = 23°; and 23° × 0·003 gives $\{\substack{\text{minus}\\-}\}$ 0·069 $\{$ (For temperature.)

And so ("Barometer corrected and reduced ") = 29·746 $\{$ (True reading.)

Let us now suppose that any person having an hour to spare every morning, and living at one fixed spot, wishes to take part in the meteorological observations so much in vogue; we may suppose, also, that he has provided himself with the best instruments that can be obtained for the moderate sum of L.10. His first inquiry, for the sake of securing barometrical accuracy, must be, "How many feet is my house above the level of the sea?" This he will learn either from the published Ordnance Survey of the district in which he lives, or by writing to one of the chief officials of this department.

This altitude once correctly ascertained, he will have no further trouble than to multiply it by 0·001, and add the product to every apparent reading of his barometer; thus, —if 35 feet above the sea, then 35 × 0·001 = 0·035; — or if 365 feet „   „   „ 365 × 0·001 = 0·365; — and the like.

If the observer's house is (say) 245 feet above the sea, and his memory very treacherous, he may write " + 0·245" on a slip of paper, and paste it on the wall, close to his barometer, for constant reference.

Thus much as to "*Correction for Altitude,*" which never varies, so long as an observer does not change his place of abode.

But the "*Reduction for Temperature*" does vary con-

8   BAROMETER.

tinually ; and therefore it is advisable to keep a full table of degrees, and their reductions, always at hand, from 32° up to 90°, never forgetting that, whilst the "Correction for Altitude" is *added* to the apparent reading, the "Reduction for Temperature" is *subtracted* from it.

In saying that the "Reduction for Temperature" is to be subtracted from the barometer's apparent reading, it should, however, be stated that this is supposing the temperature of the barometer-room to be *above* 32° (Fahr.), as it almost always will be.

Reduction for Temperature.

28°, 29°, 30° 31°, or 32° } requires ± 0·000	63° requires . . − 0·093
33° requires . . − 0·003	64 ,, . . . 0·096
34 ,, . . . 0·006	65 ,, . . . 0·099
35 ,, . . . 0·009	66 ,, . . . 0·102
36 ,, . . . 0·012	67 ,, . . . 0·105
37 ,, . . . 0·015	68 ,, . . . 0·108
38 ,, . . . 0·018	69 ,, . . . 0·111
39 ,, . . . 0·021	70 ,, . . . 0·114
40 ,, . . . 0·024	71 ,, . . . 0·117
41 ,, . . . 0·027	72 ,, . . . 0·120
42 ,, . . . 0·030	73 ,, . . . 0·123
43 ,, . . . 0·033	74 ,, . . . 0·126
44 ,, . . . 0·036	75 ,, . . . 0·129
45 ,, . . . 0·039	76 ,, . . . 0·132
46 ,, . . . 0·042	77 ,, . . . 0·135
47 ,, . . . 0·045	78 ,, . . . 0·138
48 ,, . . . 0·048	79 ,, . . . 0·141
49 ,, . . . 0·051	80 ,, . . . 0·144
50 ,, . . . 0·054	81 ,, . . . 0·147
51 ,, . . . 0·057	82 ,, . . . 0·150
52 ,, . . . 0·060	83 ,, . . . 0·153
53 ,, . . . 0·063	84 ,, . . . 0·156
54 ,, . . . 0·066	85 ,, . . . 0·159
55 ,, . . . 0·069	86 ,, . . . 0·162
56 ,, . . . 0·072	87 ,, . . . 0·165
57 ,, . . . 0·075	88 ,, . . . 0·168
58 ,, . . . 0·078	89 ,, . . . 0·171
59 ,, . . . 0·081	90 ,, . . . 0·174
60 ,, . . . 0·084	91 ,, . . . 0·177
61 ,, . . . 0·087	92 ,, . . . 0·180
62 ,, . . . 0·090	93 ,, . . . 0·183

But supposing that the temperature should ever be below 28°, the opposite process must be adopted, *i.e.*, 0·003 must be *added* to the apparent reading of the barometer, for every degree *below* 28° (Fahr.)

It may perhaps be as well to repeat that the temperature referred to, as requiring allowance to be made, is *not* that of the air outside the house, but of the air in the room, where the barometer is hung.

A good thermometer should either be attached to the barometer, or hung close to it, so that the temperature may be carefully noted at the moment of reading off the barometrical column, and due allowance made, in accordance with the foregoing table. The necessity for making this allowance may be shown by taking an extreme case : Suppose a person has two barometers, exactly equal when in the same room, and that he places one in a room artificially heated, and the other in a very cold room (when the hot room may represent an unusually hot summer's day, and the cold room may stand for the temperature of a winter's day)— the one at 90°, and the other at 35°,—

Then, apparent reading in the room at 90° is	30·200
(Reduction for temperature)	0·174
(True reading is)	30·026
And, apparent reading in the room at 35° is	30·035
(Reduction for temperature)	0·009
(True reading is)	30·026

So that between the two rooms (30·200 and 30·035) there is a difference of 0·165, or about one-sixth of an inch, due entirely to the expansive action of the heated air.

But it may be that the observer does not care particularly for such precise accuracy in his readings of the barometer, as is implied in the corrections and reductions which have been described, and yet he would like to know how much he should allow for elevation and temperature, upon an average, throughout the year. Then let us say that he is living 250 feet above sea-level, and that 9 A.M. is his favourite hour for examining the barometer, or just immediately before breakfast. To be at all nearly approaching the true reading, he must always *add* to the apparent height of the column—

In the summer months,	0·16
In the winter months,	0·24
In spring and autumn months	0·20

or, more roughly still, 0·20 all the year round.

The Standard Barometer.—For trustworthy observations the barometer should be a standard instrument, and not one of the dial-faced kind, which works a long arm by the revolution of a small wheel, thus multiplying every error. Even if the wheel-barometer (dial-faced) be a very superior one, still its *minimum* of error will amount to as much as 0·25, or a quarter of an inch both ways, that is, when very high and when very low. For example, when its true reading would be 28·50, it will stand at 28·25 (0·25 too low); or, when it ought to rest at 30·50, it will have mounted to 30·75 (0·25 too high). And this is a small amount of error compared with what many of the supposed good wheel barometers will exhibit. Therefore, always use a *standard* barometer.

It may not be amiss to return, for a moment, to the idea that was hinted at in the first pages on Atmosphere, viz., that the reason the mercurial column rises in the tube to the mean height of about thirty inches is that the whole weight of a column of air, reaching from the ground perpendicularly to the utmost limits of the air, amounts to the same as the weight of the mercury suspended in the barometric tube. Thus, imagine the tube to be square instead of round—say an inch square—the thirty inches of mercury would then be thirty cubic inches, and would be exactly balanced by the column of air, whose dimensions were cubic inches, but of unknown number, because the height of the aërial column is unknown. Well, then, as thirty cubic inches of mercury weigh fifteen pounds, so the atmosphere is said to press with fifteen pounds weight on every square inch at the surface of the earth. In the case of a column of *water,* you may consider the height to be feet instead of inches, the diameter of the column being still an inch; so that the atmospheric column, an inch square at the base, will support a water column, in a vacuum tube, an inch square and thirty feet in height. This is not the exact amount (it should be thirty-two and a-half feet), but it is near enough to the mark to show the principle on which the pump-maker works, who knows that he can raise the water from the well by a common pump, only just such a height, as that the weight of water, between the well's surface and the pump bucket, will be balanced by the pressure of the atmosphere.

Diameter of the Barometer Tube.—This should not be less than three-tenths of an inch, internal measurement; but four, or even five-tenths, or half an-inch, will be still better; for the wider the tube, the less will be the friction of the mercury against the sides of the glass.

For such a very nice observation as that for taking a mountain height, correction is made for this friction, and is called " correction for capillarity ; " it requires a small addition to the apparent height of the barometric column ; but we need not take it into account for climatic observations.

Testing the Barometer.—Before fixing the barometer to the wall, hold it securely with both hands, and gently and slowly move it sideways into a partially horizontal position.

In this way the mercurial column will run quickly to the top of the tube, and should give a clearly-heard sharp tap against the head of the tube, provided the vacuum be quite perfect. If this clear tapping sound is not produced, there must be air in the glass tube above the mercury, and this, if not removed, will prevent the instrument from acting truly.

To remove the air, the observer must slowly and carefully turn the barometer upside down; and whilst holding it thus inverted, he must continue, for three or four minutes, to tap the tube and framework with his fingers, so as to dislodge the air, and make it pass out of the tube into the cistern. Then turn it round again slowly as before, and see if the sharp tap can be heard against the top of the tube. If it still sounds dull, or not at all, it will be wise to make no more experiments, but to put it into the hands of the maker, or of an optician who understands the details of the barometer professionally. The whole operation of turning and tapping must be done very carefully; because, if the glass tube be a wide bore, as in a good barometer, any sudden change of position may cause the weight of the moving mercury to shatter the glass.

How and where to hang the Barometer.—See that it hangs perpendicularly, by applying a plumbline, and at such a height that the 30 inches on the scale may be level with the eye. Let the room be a cool one—as little subject as possible to any rapid changes of temperature. Never let the sun shine upon the instrument; and therefore, where it can be done, it will be well to choose a room with the window looking towards the north. If it can be avoided, do not hang the barometer on the *outer* main wall of the room, on account of the vibration likely to accompany windy weather.

The Vernier and its use.—The standard barometer is furnished with a permanent vertical scale, divided into inches and tenths of an inch; whilst the further subdivision into hundredths of an inch being scarcely possible, the vernier is made to obviate the difficulty. It is an ingenious instrument, intended to give the value of fractions which fall between two of the smallest linear divisions already supplied. The vernier is sometimes one inch and one-tenth of an inch in length, making eleven-tenths; and it is divided into ten equal parts, each being a tenth and a hundredth (or $\frac{11}{100}$) of an inch. It is made to slide easily up and down the permanent scale at the side of the barometer tube; but by means of a spring it is kept from slipping down from any point to which it is brought.

In taking an observation, we can, of course, read off hand the number of inches and tenths of an inch at which the mercury stands, for they are legibly marked on the permanent scale; and to find the number of hundredths of an inch, if there be any, we raise or depress the vernier, until its uppermost line, or projecting point, is exactly on a level with the surface of the mercurial column. We then count downwards the number of lines, or parts, on the vernier, which occur before we reach a line on the said vernier coincident with a line on the permanent scale; and that number of parts on the vernier represents the additional hundredths of an inch about which we were at first in doubt. It is not, however, essential that the vernier should be $\frac{11}{10}$ths of an inch in length, and divided into ten equal parts; for it would serve equally well if it were $\frac{9}{10}$ths of an inch in length, and divided into ten equal parts. In either case, the subdivision into hundredths of an inch is equally well attained. But the rule generally followed in the construction of a vernier—whether for a barometer or for circles and quadrants—is this, that the number of parts on the vernier should be equal to the denominator of the fraction, which expresses the required subdivision, whilst the value of each of its standard divisions of measure should be one more or one less. This will no doubt sound hopelessly unintelligible to any one who has never examined and worked the vernier; but a very little practice with it will explain its principle better than any amount of words.

Manner of taking an Observation.—Give the barometer a few gentle taps with the fingers, so as to help the mercury to run freely up or down the tube to its true level; but do not shake it roughly, nor be continually tapping it at short intervals.

This is not meant by way of deprecating hourly observations, under some such special circumstances as a rapid and continuous fall, forewarning the observer of the approach of a severe tempest.

For more precise information as to atmospheric pressure, and the precautions necessary in taking a delicate barometric measurement at great elevations, Mr Glaisher's "Tables of Barometrical Corrections," and other works published under the sanction of the British Meteorological Society, may be referred to with advantage.

The Indications of the Barometer.—The barometer does *not* tell us, at all certainly, whether rainy or dry weather may be expected, but, primarily, whether wind or calm is approaching. With a very high barometer we may have rain or fog; and with an exceedingly low barometer, the sky may be dry and clear. If the barometer be already high and steady, and a further rise takes place, it is often followed quickly by rain or snow. When the barometer is very low indeed, we know that a severe gale is at work *somewhere*, though it may not touch *us*. It is generally after the barometer has risen considerably from a very low point that the heavy rains take place. The greatest pressure of wind to the square inch, in a storm, is almost always recorded by the anemometer or "force of wind measurer," *after* the barometer has begun to rise. In winter, severe though not continuous frost, with or without snow, often follows close after a very low barometer.

Hourly Fluctuations of the Barometer.—It should be remembered that the barometer has, in all places, regular hourly fluctuations. From mid-day, it falls until between three and five in the afternoon ; it then rises till about ten in the evening. It falls again to a second minimum about four in the morning, and attains a second maximum about ten in the morning. The hours vary a little in different countries. But in this country we may assign ten as the morning and evening maximum, and four as the morning and evening minimum height of the mercurial column.

At the equator this fluctuation averages a tenth of an inch per day throughout the year. But in this latitude it may be said to be not much more than 0·01 of an inch per day. Still it is very perceptible; and the fact is worth bearing in mind, for this reason, that when the movement of the mercurial column is *contrary* to the regular course of its fluctuation, it is more expressive of change than when it agrees with the regular diurnal course. For instance, if the mercury fall ever so little, between 8 A.M. and mid-day, it indicates decreasing pressure far more surely than even a greater fall would, which occurred between mid-day and 4 P.M. These hourly fluctuations are, doubtless, connected with the variations of temperature and amount of vapour, arising from the sun's daily course, causing inequalities in the height of the atmosphere. For it has been found that these hours of maximum and minimum, in the daily fluctuation of the mercurial column, farther north than latitude 65°, are exactly the reverse of what they are between the tropics.

TEMPERATURE.

If we dig deep into the ground at any place, we shall find, by the insertion of a thermometer, that as we successively descend we approach constantly to some limiting temperature, which, under a certain depth, will continue unchanged. The point of this equilibrium in temperature varies in different soils; but it very rarely exceeds fifty feet. If the excavation be made about the end of November, the temperature will increase gradually as we descend in the lower strata to the limiting point; but, on the contrary, if the pit be formed in May, the ground will be found to grow colder as we descend.

In the severe climate of that region to the north of Canada, known as the Hudson's Bay Territory, it is stated by the residents there, that even in the summer months, when digging through the ground to make a grave, they always come, at the depth of a few feet, to a stratum of frozen earth. Hence it is manifest that the mass of the earth transmits very slowly indeed the impressions of heat or cold received at its surface. The external temperature of any given day will probably take nearly a month to penetrate one foot into the ground, where it is not absolutely saturated with water. By digging downwards in summer, therefore, we soon reach the impressions of the preceding spring and winter; but a farther progress downwards into the ground brings us to the temperature of the previous autumn and summer; and still lower down, we find all the various fluctuations of heat, belonging to the different seasons, intermingled and confounded in one common mean. We may therefore conclude that the temperature of the ground, at a depth of thirty, or forty, or fifty feet, is always the mean result of the impressions made at the surface during a long series of years; in other words, that whatever be the temperature of the ground at that depth, the same will be the mean temperature of the air at the surface over a number of years.

The celebrated fountain of Vaucluse, situated in the latitude of 43° 55′, and 360 feet above the level of the Mediterranean Sea, has been observed to reach its highest temperature in September, and its lowest some time in April—the former being 56°·3, and the latter 54°·1, by Fahrenheit's scale, which gives 55°·2 for its mean heat; and this corresponds very nearly indeed with the mean temperature of the place. Hence we may conclude, that however the weather may have varied from year to year, or changed its character at intervals of short periods of years, it has yet undergone no material alteration during the lapse of many ages.

This slow conducting quality of the ground, if not regulated by extraneous influence, would, doubtless, fix the heat where it was received from the solar rays, and cause such an accumulation of heat there as would gradually become insupportable. It is here that the mobility of the atmosphere plays such a very conspicuous part, as the great regulator of the system, dispensing moderate warmth, and attempering the extremities of heat and cold over the surface of the globe. As the heat accumulates within the tropics, it draws in currents of cooler air from the Northern and Southern Poles. And the activity of the winds thus produced, being proportional to their exciting cause, must prevent the heat from ever surpassing certain limits. A perpetual commerce of heat between the poles and the equator is thus maintained by the continuous agency of opposite atmospheric currents. These currents often have their direction modified; but they will still produce the same effects, only a little more slowly, by pursuing an oblique or devious course. For the actual phenomena of climate only require the various winds throughout the year to advance southwards or northwards, at the mean rate of about two miles an hour, and they will, in effect, perform three journeys of transfer annually from the equator to either pole.

Not, observe, that these currents carry the impressions of heat or cold *directly* from one extremity of the globe to another, but by their incessant play they contribute, in the succession of ages, to spread them gradually over the intervening space.

These deductions are confirmed by the results of the nicest astronomical observations. Any change in the temperature of our globe would occasion a corresponding change of volume, and consequently an alteration in the momentum of the revolving mass. Thus, if from the accession of heat the earth had gained only the millionth part of linear expansion, it would have required an increase of five times proportionally more momentum to maintain the same rate of rotation. And, on this supposition, the diurnal revolution would have been retarded at the rate of three seconds in a week. But the length of the day has certainly not experienced a variation of even two seconds in a year, since the age of Hipparchus (B.C. 140); for we cannot imagine that the ancient observations of eclipses could ever deviate an hour from the truth; and this would be about the amount of deviation in 2000 years, supposing two seconds to be the amount of deviation in each year. So that we may well believe that, at the end of 2000 years, the mass of our globe has not acquired even the ten-millionth part of expansion,—an effect which would have been brought about by the smallest fraction of a degree of extra heat. And conversely, by the same argument, we cannot suppose that the earth has experienced any, the slightest, decrease of its mean temperature.

But though there be this general uniformity of temperature at any one place, when we take the mean of centuries, yet the difference is very great between two days in different parts of the same year. Even in the mild climate of Great Britain, the thermometer will range from 92°, or even 95°, down to —12°, or 44° below freezing, making a difference of more than 100°(Fahr.) And as it is upon this excess or defect, above or below the mean temperature, that the violence of our gales mainly depends, the subject is a study possessing more than a curious interest, affecting, as it does, the commerce of a great nation, and the safety of so many precious lives, occupied in traversing the treacherous deep ; and to the farmer and amateur gardener the wisdom taught by thermometer, as well as barometer, is surely worth the pains of acquiring, when the outlay is little more than an outlay of patience,—to record the changes, as they occur from day to day, and the mode in which dame Nature always contrives to recover her lost ground, sometimes creeping up like the tortoise, sometimes bounding forward like the hare.

THERMOMETER.

Mr Lowe tells us that " the first object of an observer should be to procure good instruments, and then to place them so that they may not be affected by any *local* circumstances, such as heat from a fire, draughts in a passage, dripping of rain, or radiation and reflection from walls adjacent, &c."

"Much time is often lost by the discovery that we have been using imperfect instruments, or good ones badly placed."

"In most cases the common cheap thermometers are utterly valueless and unreliable as meteorological instruments. Mr Hartnup, of the Liverpool Observatory, mentions that among the thermometers used by the captains of the merchant service, an error of 4° or 5° is quite common ; even a thermometer fitted up for taking the temperature of water at different depths, and professing to have been made with care, was found to be more than 8° in error at one part of its scale." *Mercurial* thermometers are by far the best.

The smaller the bulb the more delicate will be the results; and the finer the bore of the tube, within certain limits, the wider will be the degree marks.

A *large* tube for the barometer, a *small* tube for the thermometer, are standing rules.

Equally in both instruments, it is essential for accuracy that the tube be equal in bore throughout, or the column will expand unequally in some parts of the scale.

The fitness of a tube for a thermometer is tested (before filling the instrument) by putting a little mercury into the empty tube, and ascertaining, with the compasses, whether this quantity of mercury occupies the same space when shifted into various parts of the tube.

If we lived in arctic regions, where the mercury is often frozen solid, we should be compelled to resort to the use of thermometers filled with alcohol, which never freezes. But where the cold seldom reaches even 10°, and never 30° or 40° below zero, it is on many accounts better to use the substance most to be relied on,—and that substance is mercury.

We can form scarcely any adequate notion of the intense severity of the frost in an arctic winter,—a severity which lasts, not for a few hours only, but for weeks together. Dr Kane, who was one of a relieving expedition in search of Sir John Franklin, says "that on the 5th of February 1854, the alcoholic thermometers indicated the terrible temperature of 75° below zero, or 107° below the freezing-point of water! At such temperatures chloric ether became solid, and carefully prepared chloroform exhibited a granular pellicle on its surface. Spirit of naptha froze at 54°, and oil of sassafras at 49°, below zero. The exhalations of insensible perspiration, from the surface of the body, invested the exposed or partially clad parts with a wreath of vapour. The air had a perceptible pungency upon inspiration, and, when breathed for any length of time, it imparted a sensation of dryness to the air-passages, inducing the men to breathe guardedly and with lips compressed." In such temperatures alcohol is the proper fluid for a thermometer, as it will not freeze, even when subjected by chemical process to a temperature of 166° below zero, or nearly 200° below the freezing-point of water.

The objections to the ordinary spirit thermometer, when it can be dispensed with, are that the spirit is apt to volatilise or float in vapour towards the vacuum end of the tube, so lessening the quantity of spirit in the bulb; this especially takes place when the temperature is high. The spirit is also less quickly affected by rapid changes of the temperature, so that it often fails to record the true maximum or minimum of the twenty-four hours. And it does not expand quite equally, for equal increments of heat at different temperatures; this, however, is only a minor objection, as its expansive action does not differ materially from that of mercury, except under a greater heat than ever occurs in climatic observations.

Supposing, then, that we determine to employ a mercurial set of thermometers,—as we cannot be continually on the watch to note the maximum and minimum of each day,—we must have two instruments, at least, which will *register*, the one the maximum and the other the minimum. Such registering instruments, if really good ones, will cost about L.1 each, or three times the price of the common registering thermometers. This extra expense will not be grudged by any one who really takes an interest in meteorology, the chief element in which is accuracy. If such registering thermometers are obtained direct from Negretti, or from Casella, in London, or from Adie, in Princes Street, Edinburgh, and if they warrant them as *tested and approved*, they may be fully relied on.

Solar and Terrestrial Radiation Thermometers.—These, again, should also register, the solar as a maximum, and the terrestrial as a minimum,—the former to be placed with a due south aspect, full in the sun's rays, the latter in any open spot, on the grass. Both these instruments should be made,—scale, frame, and all,—of glass or porcelain, which will not swell with damp, nor lose their scale-marks, as those do which have their scales engraved on wood.

Wet and Dry Bulb Thermometers.—These belong to the hygrometer, and will be explained later on, under the head of " Hygrometer Dew-point."

Rules and examples may now be given for finding—

(I.) The Mean of the Day's Temperature.
(II.) The Amount of Solar Radiation.
(III.) The Amount of Terrestrial Radiation.

To find the *Day's Mean Temperature.*—Add together the shaded maximum and shaded minimum, and divide the sum by 2, thus—

$$\text{Max. } 45° + \text{min. } 31° = \frac{76}{2} = 38° \text{ (Day's mean)};$$

or, $$\text{Max. } 58° + \text{min. } 35° = \frac{93}{2} = 46°·5 \text{ (Day's mean)}.$$

To find the *Amount of Solar Radiation.*—Subtract the maximum in the shade from the maximum in the sun; then the amount of difference will be the amount of solar radiation, thus—

Solar max. 85° − shaded max. 47° = 38°;
∴ 38° is the amount of solar radiation.

THERMOMETER. 23

To find the *Amount of Terrestrial Radiation.*—Let there be one minimum on the grass *not* shaded, and another about 4 or 5 feet above the ground, and in the shade; subtract the minimum on the grass from the minimum in the shade, thus,—

Shaded min., 28°—grass min., 21°=7° (terrest. radiation)
or,   „   31°—   „   30°=1°   „   „
or,   „   27°—   „   27°=0°   „   „

If the night be very cloudy, the terrestrial radiation will be little and sometimes nothing.

We must, therefore, have seven thermometers for all the observations, viz., for the barometer-room—for the maximum in the sun—for the maximum in the shade—for the minimum in the shade—for the minimum on the grass—and two combined, as dry and wet bulbs, for the hygrometer. All but the two last-named should be self-registering.

The only thermometers that ought ever to be exposed, either to the sun or the rain, are those for finding the maximum in the sun, and the minimum on the grass; this brings us to consider—

The best Position for the Thermometers.—The shaded maximum and minimum thermometers, for finding the daily mean temperature, and the wet and dry bulbs (hygrometer), should occupy the same spot, about four or five feet above the ground, and facing due north. The north side of a high strong wall, running east and west, will be a good position. First of all, a broad wooden board should be firmly secured to the stone or brick wall, as a kind of *lining* to the wall, and having small projecting screws, on which to hang the instruments. Next, to keep off rain, a kind of roof should be placed a foot or more above the instruments, slanting outwards and downwards from the wall. Then, again, projecting sides of wood should be placed, east and west, to keep off the rising and setting summer's sun, and they should be bored with a dozen holes or more, of half an inch diameter, for the passage of air.

The whole of this wooden enclosure should be very firm, so as to prevent vibration in stormy weather; and, at the same time, so contrived that there may be *free ventilation*. It will preserve the wood, and be less unsightly, if the whole is painted white. A small button of wood is also desirable, to press the lower part of each instrument to the wooden frame at its back; this will prevent any shifting of the maximum and minimum in windy weather.

The *Solar-radiation* thermometer should be blackened, as to its bulb, with lamp black, so as to absorb the sun's rays. It may be hung just outside a south window, with its face turned to the full action of the mid-day and 2 P.M. sun. It should be mercurial and self-registering; but the ordinary maximum will not do, unless its scale reaches higher than 130°, as the intense heat of a July sun may burst the glass, in the effort of the mercury to expand above 130°. The observation of this solar glass, for each day, may be taken any time towards evening, so as to be entered in the Meteorological Day-Book, in a column close beside the column for the minimum on the grass, which latter should be noted at 9 A.M. of that same day.

Never change the Position of the Instruments.—When once satisfied that the position is a tolerably good one, make no change. It is astonishing what a difference may be produced by this shifting from one set of local circumstances to another—enough to upset entirely any exact comparison between one day, or even one week, and another.

Daily Hour of Observation.—Excepting the solar maximum, every instrument should be examined and recorded at 9 A.M., viz., the barometer, thermometer, hygrometer, and rain-gauge. A second observation of the barometer, at 10 P.M., will give the day's true mean more nearly; but it is not essential. The observer should be very punctual to the hour of 9 A.M., and not indifferently half an-hour earlier or later. This hour (may we call it breakfast-time?) is the one least likely to be interfered with by the avocations of the day or absence from home; it is the hour by which, at the latest, every well-behaved member of society, of every rank, will be (health permitting) awake and visible. Those who think it too early to be out of their bedroom at 9 A.M. all the year round, had better give up meteorology as a study, or even as a pastime.

Manner of Observing.—In reading off the thermometers be speedy, so that the human body may not warm them; and be careful not to breathe on them. The observer will find it convenient to make up a small manuscript book of twenty pages, and rule them for the entries belonging to each instrument. This will serve to take down the observations of ten days or more, on each page, roughly, but accurately and quickly; and they can, at the observer's leisure, be copied neatly into the general day-book of larger size, which he keeps for the daily, weekly, and monthly means of many years, and which should be a large quarto, strongly bound.*

Above and below Zero.—The thermometer sinks below zero very rarely in Great Britain, perhaps once or twice in seven or eight years. Above zero may be written down as + (plus), or the sign may be omitted, as it generally is: thus + 7°, or 7° means seven degrees *above* 0°, or zero. But if you wish to record a point *below* zero, the sign —

---

* Forms for recording the various observations,—each sheet to serve for a whole month,—may be obtained for a few shillings from Casella, and after the observer has had a year's practice with them, he will be able to rule his own after the same manner.

(minus) *must* be prefixed, or it will be mistaken for a point above zero.

Decimal notation.—In all calculations, give the fractional parts in decimals, thus: Never write $29\frac{1}{2}$, or $29\frac{3}{4}$ for the barometer, but 29·50, or 29·75; and similarly for the thermometer, 44°·5, and not $44°\frac{1}{2}$. Yet, as a rule, it will be simpler to omit the decimal, in recording the thermometric maximum or minimum, and simply record the last whole degree, which the mercury has *just passed or reached*, at the same time giving the decimal for the *mean* of the two, which will thus be invariably half a degree, when any. This plan will be found sufficiently accurate for the mean of a whole week, though it may be a little above that of one day and below that of another. For example, say that the real maximum and minimum of the day are 47°·4 and 36°·3, we may state them as 47° and 37°, whose mean is 42°. Or, again, 49°·6 and 37°·2 we may call 49° and 38°, whose mean is 43°·5. So that there will never be any fraction in the stating of the maximum and minimum; and the fraction, in the statement of the mean, will always be ·5, if any at all. It will keep the book-columns of maximum and minimum more clear, and there will be less chance of error in striking the mean of the two; whilst any number of fives, in the column of the day's mean, can be run up very quickly. This, however, is all a matter of fancy, if the observer feels that he is in *no* danger of making mistakes!

Occasional testing of the thermometers.—" From time to time thermometers should be compared with a *standard* instrument of acknowledged accuracy, in order to see if the scale mark of freezing is correct; because it is well known that the zero point moves forward after high summer temperature." Mr Lowe and others assign a different cause for this shifting of the zero point, and, therefore, of all the marks on the scale. They suspect that it is not due to thermal expansion of the mercury, but to a slight contraction of the tube and bulb, under continuous atmospheric pressure, and they recommend that the thermometers should remain for a few months (or years?) filled, but not graduated. Whatever be the cause, the remedy may be found by taking off the tube and adjusting it afresh, a degree or so below its former position. But it is not worth while to do this, unless the instrument be, in other respects, a particularly good one. It would be less trouble to purchase another, and discard the offender.

Care to be taken for accuracy's sake.—Before adding up a week's or a month's column of figures, look it through very carefully, to see that the figures are in no case transposed, that is, the maximum item put into the minimum column, or *vice versâ*,—a blunder by no means uncommon, and which the eye will detect afterwards without difficulty. In the same way, the barometer's mean of each day should be examined, before adding up to strike the week's mean; for it not unfrequently happens that, in an absent mood, the observer writes, say, 29·47 for 28·47, or for 30·47, perhaps because 29 is the number of inches he is most in the habit of recording. This last warning is given under the supposition that the observer records the barometer *twice* a-day (at 9 A.M. and at 10 P.M.), when a daily mean has to be struck. If he only records once a-day (at 9 A.M.), he will have no such power of correcting an oversight.

INVENTORS, MAKERS, AND DESCRIPTIONS.—

Rutherford's Mercurial Maximum Thermometer.—In this instrument a small steel pin is pushed forward by the mercury to its highest point, and left there. It is not necessary to say much about this instrument, which is now quite superseded by others on a better principle. Its great drawback is that the pin is very liable to be entangled in the mercury by a slight oxidizing of the fluid, when it becomes utterly useless.

Negretti's Mercurial Patent Maximum.—This has no registering pin. A small piece of enamel is pushed into the tube, and secured a little distance from the bulb; if the thermometer act at once, it will always act correctly. As the temperature rises, the mercury flows over this partial obstruction of the enamel to its highest point; but it cannot return, the enamel cutting off the return passage; and so it remains to record the maximum, until the observer has seen it. By turning the bulb end of the instrument downwards, and giving it a gentle shake, all the obstructed mercury will flow over the obstruction into the bulb, or towards the bulb, ready for a future observation.

Improvement on the above.—Negretti has further improved on this idea by dispensing with the enamel and flattening the tube, at the bend near the bulb, until it is so contracted that, though the mercury will flow forward with the pressure of expansion, the mere cohesion of the mercury is insufficient to cause it to flow backwards; and so it stands at the highest point reached. As in the former instrument by Negretti, to invert it and give it a slight shake is enough to set it again for another time. It may be regarded as perfection.

Phillips' Mercurial Maximum.—A small bubble of air detaches a portion of the mercurial column, leaving the detached portion to mark the maximum. It is a valuable instrument, but scarcely equal to Negretti's last invention.

Casella's Mercurial Maximum.—This instrument is constructed on Phillips' principle, with the air-bubble to detach a portion of the mercury. It is sometimes found that these air-bubble thermometers get deranged in travelling, —the constant jolting, no doubt, moving the bubble out of its position. But supposing no such accident has happened in the transfer of the instrument from the maker to the observer, it is still constantly liable to an error of small amount from the following cause :—The air-bubble itself will expand and contract at high and low temperatures; and therefore, though it may be absolutely correct at one given temperature (say at 32°) it will read too high and too low at 70°, and at zero, respectively. Happily the error will scarcely ever amount to more than a single degree either way. All Casella's thermometers have this advantage, that they are mounted on small porcelain slabs, which preserves them from swelling like wood, or expanding and contracting like metal. This is a most important consideration, in the case of the minimum to be placed on the grass, and the solar maximum. But the most remarkable instrument for thermal purposes is *Casella's Mercurial Minimum*, which will be noticed presently at some length.

For a long time, it seemed quite impossible to construct a thoroughly good *mercurial minimum;* and, consequently, the alcoholic minimum has hitherto been in most general use, the chief objections to which have been alluded to already. But very lately Negretti and Casella have each of them offered a mercurial minimum which professes to act perfectly.

Negretti's New Mercurial Minimum retains the old steel pin, but, being pointed, it penetrates a short distance into the mercury, and, adhering to the mercury, falls with it to the lowest point to be indicated. When the temperature rises, the mercury flows past the steel pin, pressing it firmly to the side of the tube, so that the minimum record remains. It is said that this invention is a valuable one ; but I have not seen it, and therefore cannot say by what means the observer is able to bring the steel pin to the surface of the mercury again, to be ready for a future observation ; possibly it is effected, as formerly, by a magnet, though the fact that the pin must often be covered by the mercurial fluid seems opposed to this idea. The old objection of entanglement of the pin, through the oxidising of the mercury, seems to have been guarded against by some process, perhaps that of enamelling.

Casella's Mercurial Patent Minimum.—The difficulty has always been to discover some means by which the mercury *itself* might be detained at the minimum point, until observed.

In various experiments conducted by Mr Casella, it occurred to him that the adhesive property of mercury for glass *in vacuo*,—together with the fact, that where a large and a small tube are united in one bulb, the fluid will rise by expansion in the larger tube, and recede by contraction in the smaller tube,—might enable him to attain the end in view. The result is the invention of probably the first instrument known to register past indications of a low temperature, without having or forming any separate index, or in which the adhesive property of mercury for glass, as a fixed point, has ever been employed. The general form and arrangements are seen in the whole sketch, and more clearly in the enlarged section of the bend and chamber which accompany this description. $d \ldots d$ is a tube with large bore, at the end of which a flat glass diaphragm is formed by the abrupt junction of the small chamber $a\,b$, the inlet to which at b is larger than the bore of the indicating tube c. The result is, that on setting the thermometer for the next observation, in the manner to be described presently, the contracting force of the mercury, in cooling, withdraws the fluid down the small indicating stem c only; whilst, on its expanding with heat, the long slender column c does not move, because the increased bulk of mercury finds for itself an easier passage, namely, through the larger bore, and into the small pear-shaped chamber a, b.

To set the instrument, place it in a horizontal position, suspended to a nail by the back plate E, and with the lower part resting on a pin or hook at F. The bulb end may now be gently raised or lowered, causing the mercury to flow slowly, until the chamber $a\ b$ is quite empty, and the bent part $d \ldots d$ is full. At this point the flow of mercury

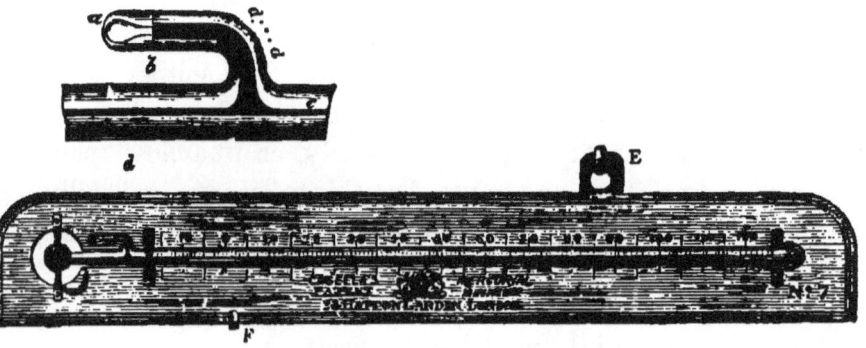

in the long slender stem c is arrested by adhesion to the diaphragm (b), and indicates the exact temperature of the bulb, or air, at the time the observer is setting the instrument. On an increase of temperature, the mercury will expand, as before explained, into the small chamber $a\ b$; and a return of cold will cause its recession *from the chamber only*, until it reaches the diaphragm b, at which point it was set. Any further diminution of heat withdraws the mercury down the slender graduated stem c, to whatever degree of cold it may attain, where it remains until further withdrawn by increased cold, or until the observer re-sets it. There is no separation of the mercury, in any of its indications, nor in the operation of setting; and the maker declares that *no changes of climate or transit can derange it.*

This mercurial minimum thermometer may be considered the most valuable addition to our already admirable set of meteorological instruments that could have been desired.

HYGROMETER DEW-POINT.

Atmospheric Moisture.—The atmosphere, in the very driest weather, is never really dry, nor at all near being quite dry. Even when the air seems to be very parching,—drying the skin, and withering vegetables,—it can be shown, by chemical agents, that aqueous moisture is present. Indeed, without it, we know that neither vegetable nor animal life could be supported. Even when the air approaches dryness very remotely, as in a close room heated by a stove and pipes, breathing becomes painfully oppressed, and there is a sensation of nervous restlessness. Whilst, if an open dish filled with water be placed on the stove, the rapid evaporation, visible or invisible, will quickly remove many of the unpleasant sensations.

Various instruments have been invented to indicate any lack or excess of the atmospheric moisture. It is on this hygrometric principle that the German toy acts, in which a miniature man comes out in wet weather, and a woman in dry. The two toys (male and female) are made to swing easily on a pivot, and the expansion or contraction of a fine piece of *catgut*, attached to them, regulates their movements. Thus the catgut is a kind of hygrometer, or rather hygroscope, indicating the presence of moisture, though not the amount. The human hair is also an indicator of moisture, as when a lady's curls droop, lank and untidy, in very damp weather.

Invisible Aqueous Vapour.—Now this moisture in the air will remain suspended as invisible vapour, so long as it does not exceed a certain quantity for a certain temperature. Raise the temperature, and the air will sustain more moisture; lower the temperature, and the air will sustain less moisture. Consequently, a hot summer day will often give absolutely more moisture than a cold winter's day; though, judging by our mere sensations, we may think the opposite to be the case.

When the air is so filled with moisture that it can sustain no more ·at that temperature, it is said to be "*saturated.*" When in this condition, let the temperature fall a few degrees (say) from 57° to 51°, then the air being unable to sustain its now superabundant moisture, some must fall to the ground, in the shape of rain or heavy dews. And if it be winter time, and the change of temperature has been (say) from 36° to 28°, then the deposit will take the form of hail, snow, or hoar-frost.

Conversely, if the air be saturated at 50°, and the temperature rises a few degrees, the air will no longer be saturated. Hence complete saturation depends on temperature as well as on the absolute amount of moisture in the air. Thus it is that a damp morning is often succeeded by a fine and comparatively dry day, the temperature rising;— and a damp afternoon is followed by rain or snow in the early part of the night, the temperature falling.

Now the mere appearance of the sky, even with a knowledge of the temperature and direction of the wind, will not tell us for certain whether rain is probable or not. What we want, to convert our guesses into something like certainty, is some instrument, or combination of instruments, which shall tell us how near the air is to a state of complete saturation at a given temperature. We shall then know, at once, what will be result of any change in the temperature.

The Hygrometer-by-Dew-point is the instrument required. This is sometimes called a psychrometer, or "cold-measurer," the fitness of which title we shall presently discover, in describing the instrument and its mode of action.

The Dew-point Hygrometer of Daniell.—If a wine bottle be brought from its cellar, it will often be found to be covered with moisture; and in proportion as the air is more or less saturated, so will the depression of the temperature (at which this moisture begins to be deposited on the cold bottle), be less or more ; that is, more moisture, then less depression of temperature, at the deposit-point ; and *vice versâ*. The degree of temperature at which moisture begins to be deposited in this way, is called the " *Dew-point,* "and hence the name of Hygrometer Dew-point given to the instrument invented by Daniell.

It consists of a doubly-bent exhausted glass tube, each end terminating in a bulb. One bulb is covered with a thin coating of gold or platinum-foil, the other with a thin linen rag. The foil-coated bulb is partly filled with ether, and holds a small thermometer, the graduated stem of which passes up the hygrometer tube, and can be read through it. If now, good ether be dropped on the rag-covered bulb, speedy evaporation ensues, and the bulb is cooled, thereby condensing the vapour of ether, which had floated into it from the ether-filled bulb, and permitting a new evolution of vapour from the ether in the bulb. This evolution of ether cools the bulb, and causes dew to be deposited on its outer foil-coated surface.

Explanation of the above experiment.—In proportion as the temperature of the air is elevated will it be capable of sustaining more moisture ; and on the contrary, by cooling the air, its power of carrying its moisture is diminished, and a portion of the moisture is condensed, on any cold body near in the form of dew-vesicles. If the air be very moist, condensation will take place quickly and largely, without any great lowering of the temperature of the foil-coated bulb.

This dew-point hygrometer of Daniell, though a great advance beyond the earlier rude experiments, is not quite satisfactory. Its construction has been improved upon by Döberemer, and by Regnault; but it still leaves much to be desired. It demands a plentiful use of *good* ether, which is expensive; and in very dry weather it is difficult to obtain the dew-deposit,—the alternate processes of evaporation and condensation not taking place rapidly enough to lower the temperature of the foil-coated bulb to the point of dew deposit.

Psychrometric Hygrometer.—This compound term, though slightly pedantic, expresses the mode of action and results obtained by it, as the " *Cold-measuring moisture-measurer.*" It consists of two good equal thermometers, mounted on the same frame of wood,—six inches wide by twelve inches long, or thereabouts,—the bulb of one thermometer being naked, whilst the bulb of the other is wrapt in some thin absorbent covering, such as a little muslin bag, with a kind of wick reaching from it into a small cistern of water, such as a small preserve-pot, or a short-necked bottle.

For the instrument to act truly, great care should be taken to choose two thermometers, which correspond exactly, degree for degree, from about 15° up to 90°. This is not at all an easy matter, for thermometers vary in the most tiresome manner, even when both are superior instruments. Both the bulbs (naked and covered, or " dry" and " wet,") should project an inch or two, clear all round, below their frame, for the action of the air to be exerted on them more perfectly. The little cistern of water should be suspended, so that the surface of the water may be from one to three inches away from the bulb, to which it is connected by the wick,—and it should be placed on the side farthest from the dry bulb, so that its evaporation may not affect the dry bulb, as well as the wet bulb.

Mr Glaisher's remarks.—" The water-vessel, or cistern, should always be supplied with rain water. If the temperature of the air (*i.e.*, the dry bulb) should have descended below 32°, it will often happen that the wet-bulb thermometer will for a time read higher than the dry bulb. *Such observations must not be recorded ;*. but when the water surrounding the wet bulb has begun to freeze, the proper readings will take place.

" In frosty weather, the water in the reservoir will be frozen ; but this is no reason for ceasing the observations, for if the water on the muslin be frozen too, the readings are perfectly available.

" If the muslin be found dry, it should be wetted with a brush or small sponge, and then be left a little time for the water on the muslin to be frozen ; and when satisfied that such is the case, the observer may proceed to take the readings in the usual way.

" Unless this caution be attended to, the wet bulb will read as high, or higher than the dry bulb. When the weather is frosty, the muslin should be thoroughly wetted some time (say an hour) before the usual and chief hour of observation ; and, as a rule, it is desirable, in frosty weather, to immerse the wet bulb and its conducting wick in water, so as to secure their having a good soaking, after every observation, ready for the next time.

" If the temperature should have ascended *above* 32° (in frosty weather) immerse the wet-bulb thermometer in warm water for a minute or so, that any ice remaining on the muslin may be melted. Unless this be attended to, the wet-bulb thermometer will continue to read 32°, so long as any ice remains in contact with it.

" The cotton lamp-wick should first be washed in a solution of carbonate of soda, and well pressed, whilst under water, throughout its length.

"The amount of water supplied to the muslin can be regulated, according as the wick is of greater or less length, from the bulb to the water; about three inches is a good medium interval between the bulb and the water surface.

"The observer must carefully abstain from breathing on the thermometers, at the moment of recording their respective readings."

The muslin and wick should so act as, by capillarity, to keep the wet bulb always wet, but *not actually dripping*, so that rapid evaporation may be constantly going on.*

The more the dry bulb is elevated in temperature above that of the wet bulb, the less is the amount of moisture in the air, in proportion to the temperature of the air, and *vice versâ*. The process of evaporation lowers the temperature of the wet bulb, beneath that of the dry bulb, either by some whole degrees, or some decimal parts of a degree.

If the two thermometers correspond exactly, when both of them are *dry*, they can never stand alike, when one of them is wetted, *except* when the atmosphere is so completely saturated, that it can take up no more moisture. In this condition of the air, the dry bulb is really as much wetted by the surrounding air, as the wet bulb is wetted by the surrounding moistened muslin. The thermometric variation of the hygrometer is therefore said to be in an inverse ratio to the amount of atmospheric moisture.

But now, supposing we find the dry bulb standing at 41°, and the wet bulb at 38°, we see that there is a difference of 3 degrees, but we cannot tell yet what is the true point of dew deposit. To know this, we must discover some formulæ or tables, which will show us that—

* The little muslin bag and its cotton wick, should be changed about once a month.

When the dry bulb is at 41°,
And the wet bulb is at 38°,
The true point of dew deposit is 34°·2.

Such tables have been constructed for us by men who have spent years in experimenting, for the purpose of devising such a *set of factors* for the dry bulb, as, when multiplied by the difference between the dry and wet bulbs, will give, as their results, the same dew-point, as that which can be found (so expensively and wearisomely) by direct observation, with Daniell's *Ethereal* Hygrometer.

Speaking of the Table of Factors, which will be given presently, Mr Glaisher says :—" The numbers in this table have been found from the combination of all the simultaneous observations of the dry and wet bulb thermometers, with Daniell's hygrometer, taken at the Royal Observatory, Greenwich, from the year 1841 to 1854, with some observations, taken at high temperatures in India, and others taken at low and medium temperatures at Toronto. The results at the same temperatures were found to be alike at these different places; and, therefore, the factors may be considered as of general application."

This method, by the use of factors, is called " Finding the dew-point by *calculation*." The other method, with ether, is called, " Finding the dew-point by *observation*."

The results obtained from the one are almost always identical with those found by the other. Let us take a specimen, such as Mr Lowe gives in his " Practical Meteorology."

By Observation. (Daniell's Hygrometer.)

Let the temperature of the air be at . .	56°·0
„ temperature of the dew-point when the ring of dew was deposited .	48 ·2
Dew-point (lower than the air)	7°·8

By Calculation (with Dry-bulb Factors).

Let the temperature of the air be as before 56°·0
„ „ of the wet bulb be . 52 ·0
Then the *calculated dew-point* is . . 48 ·2

Subtract 48°·2 from 56°·0, and the difference is 7°·8

Thus, there is the same result by both methods.

The expression "Calculated dew-point," will not be understood by any unlearned reader, until he has mastered the whole plan of factors, which he will find given at length after a few more sentences, when he may return to this point, and see that all is clear.

The credit of this idea of factors for the dry bulb belongs to Mr Belville, and was first made known to the meteorological world in the "Greenwich Observations for 1844." The table which follows is abridged from Mr Glaisher's only in so far as I have given his tenth parts of a degree, and omitted his hundredths.

Rule for using the Table of Factors.—"Multiply the difference between the dry and wet bulbs, by the factor assigned in the table to the dry bulb; subtract this product from the dry-bulb temperature; the remainder will be the temperature of the dew-point."

For example, when the dry bulb stands at 51°, and the wet at 44°, their difference is 7°; and the factor for 51° being 2°·0; then 7 × 2°·0 = 14°·0; now, subtract 14°·0 from 51°, there remains 37°.

Therefore,
When the dry bulb is at . . . 51°
And the wet bulb . . . . 44°
The temperature of dew-point is . . 37°

TABLE OF FACTORS FOR THE DRY BULB.

"*Factors by which it is necessary to Multiply the Excess of the Reading of the Dry Thermometer over that of the Moist, in order to find the Excess of the Temperature of the Air above that of the Dew-point—from* 10° (*Fahr.*) *upwards.*"

Dry Bulb.	Factors.	Dry Bulb.	Factors.	Dry Bulb.	Factors.	Dry Bulb.	Factors.	Dry Bulb.	Factors.	Dry Bulb.	Factors.
10°	8·8	25°	6·5	40°	2·3	55°	1·9	70°	1·7	85°	1·6
11	8·8	26	6·1	41	2·2	56	1·9	71	1·7	86	1·6
12	8·8	27	5·6	42	2·2	57	1·9	72	1·7	87	1·6
13	8·8	28	5·1	43	2·2	58	1·9	73	1·7	88	1·6
14	8·7	29	4·6	44	2·2	59	1·9	74	1·7	89	1·6
15	8·7	30	4·1	45	2·1	60	1·9	75	1·7	90	1·6
16	8·7	31	3·7	46	2·1	61	1·8	76	1·7	91	1·6
17	8·6	32	3·3	47	2·1	62	1·8	77	1·7	92	1·6
18	8·5	33	3·0	48	2·1	63	1·8	78	1·7	93	1·6
19	8·3	34	2·7	49	2·1	64	1·8	79	1·7	94	1·6
20	8·1	35	2·6	50	2·1	65	1·8	80	1·7	95	1·5
21	7·9	36	2·5	51	2·0	66	1·8	81	1·7	96	1·5
22	7·6	37	2·4	52	2·0	67	1·8	82	1·6	97	1·5
23	7·3	38	2·3	53	2·0	68	1·8	83	1·6	98	1·5
24	6·9	39	2·3	54	2·0	69	1·8	84	1·6	99	1·5

This method of finding the dew-point, by calculation, is speedy and inexpensive (for what can be cheaper than wicks and water?); and, if the instrument is broken, the observer's own ingenuity can quickly supply him with another, at least as a temporary substitute, until he can replace it by one entirely to his taste.

Diurnal Variation of Atmospheric Moisture.—The amount of moisture, present in the air, varies at different times in the day. It attains a maximum twice, and a minimum twice, in the 24 hours, taking one day with another. The first maximum occurs about 9 A.M., the second about 9 P.M. The first minimum is shortly before sunset, the second about 4 A.M. Popularly, the air is said to be most damp about sunrise; and in the sense of dew, or *palpable moisture*, the idea is correct, because, the temperature being very low at that hour, there is rapid condensation. But at a much hotter period of the day there may be *absolutely* more moisture than at sunrise; for the higher temperature causes it to be retained invisibly, and without imparting any sensation of moisture.

Mr Glaisher says, of the hygrometer bulbs, that " the simple inspection of the two thermometers will often afford a better criterion of the weather, and of the probability of rain, than the barometer itself; regard, however, being had to the time of the day, and of the year, when the observation is made.

" In summer, when the diurnal range of temperature is great, if in the early morning the difference between the air temperature and dew-point temperature be small, and the rise of temperature during the day be considerable, it is probable that the difference will increase; and if the temperature of the dew-point at the same time *decrease*, it is an indication of very fine weather.

" If, on the contrary, the temperature of both should increase, as the day advances, in nearly equal proportion, rain will almost certainly follow, as the air cools with the declining sun.

" In winter, when the diurnal range of temperature is small, the indication of the weather is shown by the increase or decrease in the temperature of the dew-point, rather than by the difference between the temperatures of the air and of the dew-point.

" In showery weather, the indications vary rapidly; and a person making observations, at short intervals, may predict the approach of a storm, particularly if he take simultaneous observations with the barometer."

Monthly Variation of Atmospheric Moisture.—"All months

in the year are not equally moist. In London, at Paris, Geneva, and Great St Bernard, the *absolute* amount of aqueous vapour is at its maximum in January, and its minimum in July; but the relative maximum, in those places, is in December, and the relative minimum in May." This dictum of the philosophers seems to need some explanation, as it implies that the temperature of December is lower than that of January, and of July lower than that of May (?).

Fluctuations of Aqueous Vapour.—It is found that the evaporation of a single day can only raise the dew-point temperature a very little way, so that it is nearly the same at noon as it was the previous night. And as its temperature at night was determined by the lowest temperature of the night air, we have the means of learning pretty nearly the dew-point for any day, by finding the lowest temperature of the previous night. Thus, if the shaded minimum gave the previous night at 36°, the mean dew-point temperature, for the whole of the next day, will be *a little above* 36°.

Similarly, the dryness of a day depends on the difference between that day's maximum and the previous night's minimum temperature:—Thus, if to-day's maximum be 60°, and last night's minimum were 40°, the day's mean dryness is very considerable. But if the night's minimum and the next day's maximum temperature be nearly equal, the air of the whole day will approach nearly to complete saturation. And so it appears that dryness depends on cold nights being followed by warm days; and, therefore, one of the most important of the thermometrical observations is the *range* of the day's temperature. It is often noticed that a warm night is followed by heavy rains; but an extremely cold night can hardly ever be succeeded by a rainy day.

Again, if we take the night's shaded minimum and the next day's *mean* temperature, the difference in amount of vapour sustainable at each will give the average dryness of the day. Suppose, for example, that the night's minimum had been 35°, and the following day's mean were 42°, then

Vapour sustainable at 42° = . . 0·283 (inches),
   „   „   35° = . . 0·222 („   ),
Therefore, dryness or capacity for more ——
moisture . . . . . = 0·061 („   ).

And now let us compare the dryness of a summer's day with that of a day in winter. We will suppose the same range of 20° in each case. Let the summer's day maximum have been 70°, and its minimum 50°, then

Vapour sustainable at 70° = . . 0·727 (inches),
   „   „   50° = . . 0·373 („   ),
(*N.B.*) Therefore, the extreme dryness of ——
the air, ... = 0·354 („   ).

Next, let the winter's day maximum have been 50°, and its minimum 30°, then

Vapour sustainable at 50° = . . 0·373 (inches),
   „   „   30° = . . 0·192 („   ),
(*N.B.*) Therefore, the extreme dryness of ——
the air, ... = 0·181 („   ).

Thus, the extreme dryness of a winter's day is only half that of a summer's day, though the range be 20° in both; and when we remember that this range of 20° is not unfrequent in summer, and is very rare, indeed, in winter, we may judge what an immense difference there is in the dryness of the two seasons. These fluctuations in the amount of aqueous vapour are therefore directly dependent on the course of the temperature. As a rule the air is drier at the increase than at the fall of the day and of the year,—the morning is drier than the evening, and the spring is drier than the autumn. In the one case, the temperature is hourly or daily increasing, and leaving the dew-point, as it were, at a greater distance behind it; in the other case, the temperature is hourly or daily falling, and coming gradually nearer and nearer to the point at which some precipitation of superabundant moisture must take place.

In his preface to the hygrometrical tables, Mr Glaisher says:—"The two thermometers should be delicate and sensitive in the extreme, such as have been made for several years by Messrs Negretti and Zambra. These opticians have enabled me, up to the present time, to carry out my idea of the requirements of good thermometers. The instruments supplied by them are graduated, with the utmost nicety, upon their own stems; and their bulbs are small, so as to insure sensitiveness.

"In my recent balloon ascents, I took a series of simultaneous readings with the dry and wet bulb thermometers and Daniell's hygrometer, for the purpose of determining to what elevation above the sea the dry and wet bulb thermometers could be used with confidence. The following are some of the results:—The temperature of the dew-point, by dry and wet bulbs, at heights less than 5000 feet, was the same as by Daniell's hygrometer, from 73 experiments; at heights between 5000 and 10,000 feet, was $0°·1$ higher than by Daniell's hygrometer, from 29 experiments.; at heights between 10,000 and 15,000 feet, was $2°·0$ higher than by Daniell's hygrometer, from 3 experiments; at heights between 15,000 and 20,000 feet, was $0°·5$ higher than by Daniell's hygrometer, from 9 experiments. The results, therefore, by Daniell's hygrometer, are identical with those by the dry and wet bulb thermometers, up to about 10,000 feet. At heights exceeding 10,000 feet, the dew-point was somewhat lower by Daniell's hygrometer than by the dry and wet bulb thermometers, but at those heights the temperature was very low.

"The general result, from all these experiments, is that the dry and wet bulb thermometers may be used with confidence up to great heights above the sea-level."

"*Its use in the sick-room.*—The value of this instrument in a sick-room will be at once obvious to those who know that the comfort of the patient often depends quite as much on the hygrometric state of the air in the room, as on its temperature. If the air in an apartment be too dry (that is to say, if the difference between the readings of the dry and wet bulbs be as much as 12°, or even more, it may be remedied by placing on a table in the room some broad, shallow vessel, such as a small tea tray, partly filled with water, when evaporation will gradually give the air a pleasant and healthy degree of moisture; or a pan of boiling water, open and steaming, will produce the same effect far more speedily. If, on the contrary, the room's atmosphere be too damp, it may be corrected either by raising the temperature considerably with a larger fire, or (if this appear to the medical men not advisable) by placing a small open vessel, partly filled with sulphuric acid, or any other medium capable of rapidly absorbing moisture.

"The dry and wet bulbs will serve as a guide either way. The bulbs should be so placed that neither the fire nor draughts may produce partial or local effects upon them. A difference of 7° or 8° between the two bulbs will be found to give a very pleasant sensation of healthy moisture.

"*Its use in greenhouses and hot-houses.*—It does not follow, because the temperature is kept at the proper height, that the air will be conducive to healthy growth; for it may be far too dry. A large shallow tank, with a sliding lid, to uncover more or less surface of water, as required, will remedy this dryness; and the dry and wet bulb thermometers will enable the gardener to regulate the moisture with a very little practice."

THE RAIN-GAUGE.

There are several constructions of this instrument; but little need be said about any of them, as they are all good in their way. If the instrument be effective, it is the better for being as cheap, and, therefore, as simple as possible. To be effective, the gauge should be an equal cylinder, whose diameter is always the same from top to bottom. It should therefore have no enlarged rim at the top, but a sharp edge all round; or otherwise the diameter, at its uppermost point, will be less than lower down, and so admit less than the true quantity of water. In theory, the edge should be as sharp as a razor, so as to *cut in two* any drops of rain that fall upon it, in order that no minute splashings from without the area of the gauge may find their way into it. And further, it should be covered in, (funnel fashion) *a few inches below* the top,—a very small hole being left in the centre of the funnel-cover, enough to permit the rain to fall through. A float, with a very slender stem graduated in inches and tenths of an inch, and passing up the aperture, will both record the rain-fall, and partially close the passage against loss by evaporation. Indeed, this arrangement will so reduce the loss by evaporation in hot weather, that we may reckon it as "*Nil.*"

The more complex gauges, of the kind provided with a graduated glass tube at their side, are handsome instruments, but far better suited to tropical climates, where frost is unknown. In this country, the frequent frosts of winter will render the tube very liable to breakage, through an occasional oversight in not emptying the gauge of its contents immediately after rain.

Mr Adie of Edinburgh supplies the Meteorological Society, and the public generally, with a very good gauge at about L.1.

Complaints are sometimes made that evaporation takes place in hot dry weather through the small funnel aperture. This need not be a serious objection, if observers would only examine their rain-gauge *every* morning at 9 A.M., and, if rain has fallen, empty it down to zero; and if they find, in very dry weather, that it is a little below zero at the end of a few days, they may raise the indicating stem to its proper place, at 0, by pouring in a dozen drops from a small watering-can. Of course, if they leave the gauge to take care of itself from week to week, perhaps even for a whole month, evaporation in summer will certainly rob it of a portion of its proper contents.

A small brass tap below allows the gauge to be emptied, after the observer has recorded it. The tap is secured from accidents by the protection of the larger cylinder, into which the whole gauge is sunk. This outer cylinder should be sunk in the grass, in any open spot, free from trees or buildings for fifteen or twenty yards clear all round, to such a depth that the rim of the gauge may be about one foot above the ground.

Those who are interested in the subject of the "British Rainfall," may receive a monthly abstract, and a full annual report, with much varied information, in return for a yearly subscription of ten shillings, by applying to F. Symons, Esq., M.D., Camden Road Villas, London.

Snow Measurement.—When a wet, sleety kind of snow

falls, melting almost as fast as it reaches the ground, the rain-gauge will give the depth of melted snow, just as if it were rain. But if the snow be of a more frosty kind, and lies to any depth, it may happen that the rain-gauge retains only a small part of it, especially if there be any wind; and, obviously, if the snow amount to six or eight inches, the ordinary rain-gauge, covered in with a funnel, as it should be, is quite useless for purposes of measurement.

The depth *in inches of water* may be found in two ways, first, by measuring the depth of fallen snow, in any open spot, with a common foot-rule, and dividing the number of inches by 8, 14, or 20, according as the snow is wet and close—dry and frosty—or very light and feathery; and, secondly, by lifting a portion of the snow, and measuring after melting it.

This method is as follows:—Take a well-made cylinder, exactly *eight inches in diameter*, and eighteen inches deep, though the depth is immaterial, provided it be deep enough. Invert it on the snow, in some level place free from drifts; press it down through the snow to the ground; if its edge be very sharp, it will not displace the snow much. Then clear away the snow all round, and pass under the cylinder a very thin broad piece of metal. You may now lift the cylinder, "snow and all," and carry it in doors, and melt the snow in the cylinder. Next, pour the water into another smaller cylinder, which is to be exactly *four inches diameter*, and *four inches deep*. (The diameter and depth must *both* be exact.)

This four-inch measure, when quite full, will contain exactly one inch of water, as collected in an eight-inch gauge or cylinder. If the melted snow does not quite fill the four-inch measure—or, having filled it once, will partly fill it again—you may obtain the number of decimal parts of an inch, by plunging to the bottom of the four-inch measure a little rule, exactly four inches long, and divided into one hundred equal parts, as in the accompanying sketch. For, as the whole depth of four inches in the little measure represents one inch of water, as rain-fall, in the larger measure, so each of these divisions on the rule represents 0·01 of an inch. If the 8-inch and 4-inch cylinders are accurately made, this method of computing the depth of fallen snow, in the form of water, may be regarded as the best.

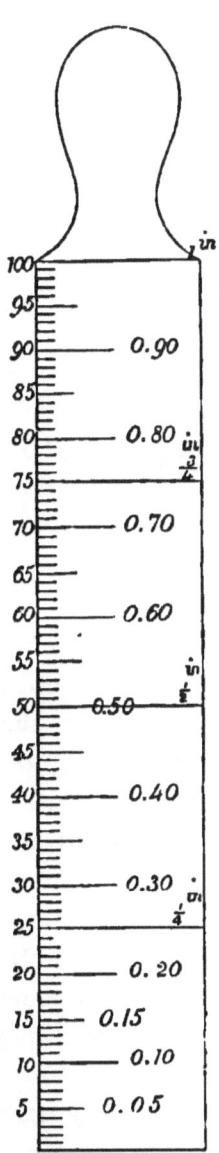

In melting the snow in the 8-inch cylinder, it will be a good plan to place the cylinder in front of the fire, with the flat piece of metal lying over the mouth of the cylinder, so as to prevent evaporation as much as possible.

If you wish to be quite satisfied that the two cylinders are accurately made, you may do so in the following manner:—Pour water into the larger cylinder to the depth of six inches exactly. If the cylinders correspond in their proportional sizes, this depth of water in the 8-inch cylinder will fill the 4-inch cylinder six times.

Variation of Rain-fall.—Of all the instruments in general use by meteorologists, there is not one in which *locality* produces such very different results, even in the same district, as in the rain-gauge. The vicinity of a mountain range adds enormously to the amount of rain-fall. For, though the mountain may occasionally

c

divert a passing storm, so that some particular fall of rain or snow may be less at the mountain's foot than at a place some miles distant from it, yet the total rain-fall of a month, and still more of a whole year, will be found to decrease regularly the farther we advance from the lofty range.

Thus, in some of the rain-gauges stationed near Loch Katrine, more than a hundred inches of rain are recorded as the year's total fall. In the district of Callander, the annual amount of rain is between sixty and seventy inches. At Muthill, near Crieff, the fall averages about forty-three inches. At Colquhalzie, at Auchterarder House, and at Trinity-Gask, it is a few inches less than at Muthill. And when we arrive at Perth, we find that the year's rain-fall there is often not much above twenty inches.

A glance at the map of Perthshire will show that this decrease of rain-fall, at the places I have named, follows, in regular steps, the increase of distance from the great centres of rain, or the lofty south-western ranges of the county.

Now, in view of such an unequal distribution of rain, the first thought that naturally occurs to the mind is, that if the land near the hills receives no more rain than is necessary for its productiveness, the lands farther away must receive too little; or, *vice versâ*. And yet it is not so—certainly not in these islands—because, by the wear and tear of ages, the richest soil, and the soil least requiring frequent showers, is farthest from the mountain side. There are some exceptions, of course, to this wide statement; but the law is a general one over the whole of those portions of our earth, known as the Temperate Zones.

On examining the soil of a mountain farm, we find it to be little else than disintegrated rock, reduced to an almost impalpable powder by the action of wind and water, aided by the constant alternations of heat and frost. And whatever of producing power may be imparted to a soil thus situated, either by the art of man, or by the natural decay of the scanty vegetation around, it is quickly carried through, downwards and forwards, to the lowest level, by the filtering of water through the porous stratum, and in obedience to the universal law of gravitation.

The result is, that the lower lands of the strath, by the alluvial depositions of many ages, have obtained by far the larger share of what we may be allowed to call nature's producing elements. To this must be added the *mechanical* effects of a low level, in that the water supplied there by rain lies in store, in a kind of subterranean basin, ready to be drawn up again by the evaporating power of the sun, for the support of the vegetable life above; whereas, in the case of the higher grounds, the frequent fresh supply of water passes rapidly away, by invisible underground channels, to the lower levels. And thus it is that our lowland farms and gardens are rich with healthy verdure, under a continuance of heat and drought which would destroy all highland vegetation to the very roots.

FACTS AND THEORIES.

Classes of Cloud.—The clouds were first classified and named descriptively by Luke Howard, the meteorologist. The names were afterwards abbreviated by Mr Glaisher thus—

Ci = cirrus, ... feathery looking, most lofty cloud.
Cu = cumulus, . . . mountainous looking.
St = stratus, ... { the ground cloud, forming at sunset, and disappearing at sun-rise.
Ci-cu = cirro-cumulus, rounded masses, or woolly tufts.
Ci-st = cirro-stratus, . horizontal masses, rather lofty.

Cu-st = cumulo-stratus, layer of cumuli, also lofty.

Ni = nimbus, . . . { the rain-cloud, often seen with the stratus.

Sc = scud, ... { a broken portion of the nimbus, smoke like, and nearly vertical.

The stratus of the night often rises into detached cumuli, in fair and warm weather; and if these (cumuli) go on increasing in number and density towards mid-day, thunder may be expected before night.

The cirrus is seen mostly in fine steady weather.

The cumulus is for the most part a summer cloud.

The cirro-cumulus betokens a dry, calm atmosphere.

The cirro-stratus is often the first visible token of the breaking up of a long continuance of settled fine weather; it then rises in the eye of the *coming* wind, and gradually overspreads the heavens with a pall, which grows denser and denser, till rain or snow falls.

The scud is very partial, often pouring its contents of rain, hail, or snow, on one field, and scarcely even sprinkling another field, within a distance of a few hundred yards.

Dark patchy masses, horizontal and detached, a little below a curtain of paler cloud, generally betoken heavy rains.

An entire absence of clouds, with distant objects standing out very sharply against the clear sky, is often the precursor of an easterly fog or drizzle.

A combination of many of the above classes of clouds is seen occasionally, at one and the same time; but in general, the day is very plainly characterised by the presence of one class of clouds, to the exclusion of all the other classes.

Thunder and Gales at the Equinox.—It often happens that, in spring and autumn, the character of the coming two or three months is determined, pretty closely, by the equinoctial disturbances, which precede them. Thus, in April a thunder storm will sometimes change the weather most materially; if coming up from a westerly or southerly quarter, by introducing moist nights and warm days; and if coming from the east, by leading the way to a long dreary succession of dry, keen, easterly winds.

As a rule, the prevailing winds of autumn are from the west, and those of spring from the east; and when this rule of the seasons is departed from to any great extent, the balance is equalised in the succeeding months.

Sometimes the weeks, immediately before and after the equinox, pass by almost entirely free from any great atmospheric disturbance. When this is the case with the autumnal equinox, the temperature continues higher than usual, until far into the winter months; and, on the contrary, a calm spring equinox maintains a very low temperature for many succeeding weeks. The reasons for this are easily perceived, when it is remembered that the one chief cause of high winds is the difference of temperature in vast masses of adjacent air. If this difference be less than usual, the result will be calm, at both the equinoxes, with higher than average temperature in the autumn or early winter months, and with lower than average temperature, in the spring and early summer months.

The Velocity of Light.—It was accepted until very lately, as an established fact, that light travels at the rate of about 192,000 miles in a second of time. But doubts are now entertained as to whether our earth's mean distance from the sun be, as was formerly thought, 95,000,000 of miles. When the true mean distance between the sun and the earth is determined, we shall, perhaps, be compelled to alter our axiom as to the velocity of light. The difference will not probably be very great—a twentieth, perhaps, more or less than we have reckoned it hitherto, or from 184,000 to 200,000 miles in a second.

The Velocity of sound.—It has been calculated that, when

the temperature (?) and pressure of the atmosphere are about at their mean, Sound travels, in an open space, with a speed of about 1120 feet in a second, or about a mile in five seconds. We may thus determine the distance at which a flash of lightning is from us, by counting the number of seconds between the flash and the sound of the thunder. Suppose twenty-five seconds have elapsed, before the thunder is heard, we may reckon that the focus of that particular flash was distant about five miles from us.

The Velocity of wind.—A light pleasant breeze moves forward at the rate of about 4 or 5 miles an hour; a fresh breeze at from 10 to 20 miles an hour; a high wind at from 30 to 40 miles an hour; an ordinary storm at about 50 or 60 miles an hour; a raging hurricane at from 80 to 120 miles in an hour.

Hurricanes are chiefly confined to tropical regions, and above all are peculiar to the West Indies.

Tornado is the term generally applied in speaking of African hurricanes.

Typhoon is the name given to the hurricanes in the seas around China and Japan.

The Symoom is the hot desert wind of Africa, Egypt, and Syria.

The Sirocco is the hot wind of Italy.

The Mistral is a cold north-eastern wind, which prevails at times in Southern France, and is said to be the most tempestuous wind with which Europe is ever visited.

The earth's rotary motion from west to east has no doubt some effect in directing the gradual, and apparently circular changes in the direction of all the winds; but thermal expansion must be considered as the chief cause of any very great aërial commotion.

The Aurora Borealis.—We often see near the northern horizon, and usually a short time after sunset, a dark segment of a circle, surrounded by a brilliant arch of white or fiery light; and this arch is again often separated into several concentric arches, leaving the dark segment visible between them. From these arches, and from the dark segment itself, in northern latitudes, columns of light, of the most beautifully variegated colours, shoot up towards the zenith; whilst irregular masses, like sheaves of light, are scattered in all directions. The appearance is then surpassingly grand, as the whole heavens seem to move, in measured cadence, to the general undulation of the masses of light.

It has been asserted that in the arctic regions, these brilliant appearances are attended with loud noises, resembling the hissing and crackling of fireworks. If this statement could be relied on, there would be no further doubt as to the distance from our earth at which these magnetic currents play to and fro, because the transmission of sound would at once show that they were in contact with our atmosphere, which acts as a medium to convey the sound to our ears. But Captain Lyon, who accompanied Parry in his arctic expedition of 1819, denies that any sound was ever heard by him during some of the most brilliant auroras that man, perhaps, ever beheld. He tells us that the sudden glare and rapid bursts of those wondrous showers of fire make it difficult to persuade oneself that their movements are wholly without sound. He declares that he stood on the ice for hours listening, and at a distance from his men and the ship, so as to secure stillness of all things around him, and he was thoroughly satisfied that not the very faintest sound came from the Aurora.

It is now generally believed that these streams of light are magnetic currents, periodically evolved by the earth itself—intensified occasionally by magnetic sympathy with certain changes taking place in the sun; and that particular conditions of the higher stratum of our atmosphere tend to make their passage through it visible to us. It is probable that though, at a vast height above the earth,—far beyond the clouds,—the line of their passage is within fifty miles of us; or, in other words, that the outer and most attenuated of our atmospheric layers is the one in which they play their merry gambols.

A similar appearance has been observed by travellers in the southern hemisphere, and proceeding from the South Pole, and therefore called "Aurora Australis." Hence it would be more appropriate to speak of these phenomena as "Polar Lights."

Thermal Expansion of Water.—The general law that bodies expand when heated, and contract when cooled, only applies to water within certain well-defined limits. Thus, down to 39°, water contracts in bulk; from 39° to 32°, it rather increases in bulk: whilst in the act of becoming ice, at or below 32°, it undergoes a *vast* expansion. By this Providential departure from a general law in nature, the ice is kept at the *surface* of the water.

If the general law of contraction in bulk, at decreasing temperatures, had prevailed in the case of water, without the marvellous exception just named, the ice would have sunk to the bottom of our rivers and lochs,—one layer after another,—and the ocean itself, except perhaps just under the equatorial line, would have become, ages ago, a solid impassable mass. Even in summer, the sun's rays would have had but little warming power over the air, in these now temperate regions, on account of the immense cooling influence which would have been exerted over all the land, by the surrounding ocean of ice.

Atmospheric Aqueous Vapour.—Professor Tyndall's recent discoveries respecting the absorption and radiation of heat by aqueous vapour, are an important addition to our previous knowledge. They are of great value as illustrating the vibratory actions in matter, which constitute heat. But it is in connection with meteorology that they chiefly command our attention. From his experiments we learn that the minute quantity of water, suspended as invisible vapour in our atmosphere, acts as a warm clothing to the earth. The efficacy of this watery vapour in arresting and preserving radiant heat is, in comparison with that of air, perfectly astounding. Although the atmosphere contains on an average, only one particle of aqueous vapour to 200 of air, yet that single particle absorbs eighty times more heat than the collective 200 particles of air, or 16,000 times more heat than that absorbed by one similar particle of air.

Professor Tyndall says that "if you were to remove, for a single summer night, the aqueous vapour from the air which overspreads these islands, you would assuredly destroy every plant which is not capable of bearing extreme cold. The warmth of our fields and gardens would pour itself unrequited into space; and the summer's sun would rise in the morning with scorching power, on an island fast bound with frost."

Mr Glaisher has settled the question, previously conjectured, that rain falls only when cloud exists in *double layer*, and that rain-drops diminish in size with elevation, merging into wet mist, and above that into an apparently dry impalpable fog. In his balloon excursions he has met with snow falling for a mile in thickness, *below rain*,—contrary to our preconceived notions,—showing that the decrease of temperature with elevation does not follow (so steadily and regularly as had been assumed) the law of about 3° diminution of heat for every thousand feet of elevation.

The Gulf Stream.—Whatever may be the urging cause of the Gulf Stream, the direction of its current and its influence upon the British Isles are plainly to be traced. Setting out from the hot regions of the Gulf of Mexico and the

Caribbean Sea, it proceeds northward to the great fishing banks of Newfoundland, and thence to the shores of Europe, yielding up its heat to the westerly winds, and thus transferring a portion of the superfluous warmth of the tropics to our colder shores.

The greatest heat of the oceanic water, in the Mexican Gulf, is about 87°, or about 9° above the ocean temperature that would be due to latitude alone; and when it has reached 10° of north latitude, the Gulf Stream is found to have lost only about two or three degrees of its original heat, and after ascending northwards a distance of three thousand miles from its starting-point, this mighty oceanic river still maintains a summer heat, when all outside of it is cold as winter. It thence crosses in a westerly direction, and spreading itself more widely, imparts to Western Europe a degree of mildness which is much greater than would be due only to latitude. Ireland and the British Isles split the stream into two directions, one tending to the North-North-East, the other passing down the west coast of France.

Every westerly and south-westerly breeze that blows towards us crosses this wonderful body of warm water, and comes floating over the lands laden with warmth and moisture, to mitigate the wintry blasts and the summer's burning sun.

If we wish to realise in imagination the full benefit of the Gulf Stream to these islands, we have only to compare our condition in winter and spring with the state of the harbours of Labrador and Newfoundland, in the same latitude. Whilst they are rigid with ice as late as May and even June, we are robed in green, and our harbours at the coast have scarcely ever known what it is to be locked with ice, even in our severest winters.

Magnetic Sympathy between the Earth and the Sun.—Mr

Nasmyth's theory (shall we say *discovery?*) is that the bright surface of the sun is composed of an aggregation of apparently solid forms, shaped like willow leaves, or some well-known forms of diatomaceæ, and interlacing one another in every direction. These forms are so regular in size and shape, that one of our profoundest philosophers suggests their being *organisms*, perhaps even partaking of the nature of life, but at all events closely connected with the heating and vivifying powers of the sun. These mysterious objects are computed to be not less than a thousand miles each in length, and one hundred miles in breadth! The enormous chasms in the sun's photosphere, to which we apply the diminutive term "spots," exhibit the extremities of these leaf-like bodies pointing inwards and downwards,—fringing the sides of the cavern, far down into the abyss, sometimes forming a kind of rope or bridge across the chasm, and seeming to adhere to each other by lateral attraction.

It now appears certain that sympathy exists between forces operating in the sun, and magnetic forces belonging to the earth. A most remarkable phenomenon, tending to prove this sympathy, was seen by different and independent observers, at two distant places, on the 1st of September 1859. A sudden outburst of light, far exceeding the brightness of the sun's surface, was seen to take place, and sweep, like a drifting cloud, across the solar disc. This was attended by *terrestrial* magnetic disturbances of very unusual intensity, and with most extraordinarily brilliant auroras. The identical instant at which the solar effusion of light took place, was recorded by an abrupt and strongly marked deflection in the self-registering instruments at Kew.

If conjecture may be hazarded, we may suppose that this remarkable event had some connection with the means by which the *sun's heat is renovated*. It is a reasonable supposition that the sun, at this time, was in the act of receiving a more than usual accession of new energy. And the theory which assigns the maintenance of the sun's power to the plunging of cosmical matter into it,—with that prodigious velocity which gravitation would impart, as the foreign body approached to actual contact with the solar orb,—would explain this sudden exhibition of intensified light, in accordance with our previous knowledge, viz., that "suddenly arrested motion results in equivalent heat." (?)

Component Parts of the Atmosphere.—The atmosphere may be treated of in various ways, according to the nature of the subject to which it bears reference. Thus we may regard it *statically*, that is, *at rest*; or *dynamically*, that is to say, *in motion*. Its relation to the imponderable agencies of heat, light, electricity, magnetism, and actinism, is another way in which to view the atmosphere. Or again, as in the present instance, we may wish to speak of its chemical conditions, and to enumerate its component parts.

The air is composed of four different bodies, which are *mixed*, but generally believed to be *not chemically combined*—each existing as an independent atmosphere, penetrating through the others to the greatest height, from which specimens for analysis have been brought, and being found in these specimens always in about the same proportions as in the air at the level of the sea.

One hundred parts of air consist mainly of 78 or 79 parts of *nitrogen gas*,—called also azote,—and about 21 parts of *oxygen gas*. The other two chief ingredients are in very small proportions, as *carbonic acid gas*, about the $\frac{1}{1200}$th part, varying to a little more or less; and *watery vapour*, also very variable in quantity, but rarely exceeding 1 part in 100 parts of common air.

The atmosphere contains also minute portions of ammonia and nitric acid, and occasionally of hydrogen, said to be most manifested during thunder-storms. It also contains in the form of vapour a multitude of adventitious substances, in those injurious mixtures known as *miasmata*, the nature of which can hardly be investigated. And in the near neighbourhood of large towns there will, doubtless, be other temporary ingredients floating about, of serious consequence to the health and comfort of the inhabitants, but of infinitesimally small amount when compared with the whole mass of the atmosphere at places distant from those crowded quarters of humanity.

Atmospheric Electricity.—The great constant fact respecting atmospheric electricity is, that the earth is always becoming *negatively* charged, while the air, or all conducting substances contained in it, is at the same time *positively* charged. In fine weather this goes on without interruption, but it is different for different hours of the day. When the state of the air is steadily observed by an insulated conducting rod, we find that it is least charged at eight in the morning, and most charged between eight and ten in the evening. In foggy weather the states are reversed, and the air is generally negative. During an actual shower the air is commonly positive, but the rain-drops themselves, when examined, are found to be in an opposite state. But during showers, and especially during thunder-storms, the state is constantly changing from one kind of electricity to the other. The dense fogs which sometimes settle over a place, and last for hours, have usually a strong negative character; and their being thus highly charged with electricity, seems to prevent them from being dissolved by heat so readily as vapours generally are, for it has been observed that a cold fog entering a warm room will remain a considerable time in the visible form. The ordinary action taking place upon the air will, therefore, render all floating clouds liable to become negatively electric, especially in the lower regions, if we suppose that it is at the surface of the earth that the electricity is generated. If other clouds float above, with a stratum of dry air between, these may become positive by induction, and thus there will be an opposition of states among the clouds themselves, as well as the opposition between the lower strata and the earth. When the accumulation has reached a certain pitch, it will be discharged in some one of the three following ways:— If there is a communication of moist air between the two charged clouds, or between a cloud and the earth, it will pass off quietly by *conduction*, with or without the deposition of moisture in the visible form of rain, hail, or snow, &c. Or, if there is a great movement among the clouds themselves, it may be neutralised by *convection*. But if two contiguous masses are highly charged, and a dielectric of very dry air lie between them, it will be discharged by *disruption* in the form of thunder and lightning. When two clouds are mutually discharged in this way, the whole commotion takes place in the upper air, the lightning being faint and diffused, and the thunder always distant. But when a large mass of positive cloud hangs over the negative earth, and when the intervening air is dry, and the electric accumulation great, the discharge bursts forth between the sky and the ground, and is then the most terrific and dangerous. The light has then the zig-zag known by the name of forked lightning, and it is apt to be concentrated in some one spot, which is torn and scorched by the stroke. The thunderbolts which throw down or set fire to houses, shiver trees, and destroy life, are of this character. All the effects capable of being produced on a small scale by electrical jars and batteries are caused on a grand scale by these thundery discharges between the clouds and the earth.

What is Ozone?—A summary of our knowledge relative to ozone may be thus briefly stated :—Oxygen gas is susceptible of undergoing a change, the nature of which is almost wholly wrapt in mystery. It is capable of becoming odorous, corrosive, and irritating when breathed. Its chemical action is capable of being so modified and increased, as that, whilst ordinary oxygen gas neither bleaches nor corrodes silver, nor decomposes iodide of potassium, the allotropic or second form of oxygen gas will accomplish all these, and many other results peculiar to itself. It will be found capable of removing the colour of sulphate of indigo, and other vegetable and animal colourings. All this is due to the modification which ordinary oxygen gas has undergone, and which is known as its *conversion into ozone*.

Electricity in the air is believed to be one of the chief agencies in the production of ozone.

Ozone is inconceivably beneficial in rapidly removing the bad odours which arise from the decomposition of animal and vegetable bodies. If the temperature be high, and very little ozone be present in the air, decomposition will take place on every side, charging the air with disease and death. But ozone, though so very useful an agent, may become injurious when present in the air to any great excess, by its irritating effects on the sensitive air passages of the throat and lungs.

As an ozonometer, the observer may obtain test papers from several makers, which may be hung in the open air, but sheltered from the sun and rain. According as ozone is present in the air, in greater or less abundance, the test-papers will be coloured with varying shades of *blue*. The intensity of these shades of blue may be measured by comparing them with a graduated scale of twelve tints, which may be obtained, together with a packet of the prepared slips of paper, from Messrs Casella, 23 Hatton Garden, London (E. C.), at a cost of 6s. 6d.

If the observer wishes to try the experiment of preparing his own test-papers, they may be made thus :—

200 parts of distilled water,
10 „ starch,
1 „ iodide of potassium,

which should be boiled together for a few seconds, and thin strips of bibulous paper should be dipped into the solution ; and when dry, they will be ready for use. If there be any ozone present in the air, it seizes on the potassium—the blue colour left being iodide of farina. It will be as well to keep the strips of prepared paper (until wanted for use) in a box or drawer, dark and dry.

Meteoric Stones, or Aërolites.—These are solid semi-metallic substances, which fall from the atmosphere. The larger stones have been seen as luminous bodies, moving with great velocity, descending in oblique directions, and frequently with a loud, hissing noise, resembling that of a mortar-shell when projected from a piece of ordnance ; they are sometimes surrounded with a flame, tapering off to a narrow stream at the hinder part, and are heard to explode, and seen to fly in pieces. Of course these appearances have been observed most frequently in the night ; but when the stones have fallen in the day-time, only the report and shower of stones have been noticed.

The same meteoric mass has often been seen over a great extent of country—in some instances, 100 miles in breadth, and 500 in length, which implies that they must have had a great elevation. Indeed, from various calculations, it appears that during the time that they are visible, their perpendicular altitude is generally from 20 to 100 miles ; and their diameter has, in some instances, been estimated to be at least half a mile. Though rarely visible for a whole minute, they are seen to traverse many degrees in the heavens. Their rate of motion cannot be, generally, less than 300 miles a minute. Probably the stones which come to us from them form but a very small portion of their bulk, while the main body holds on its way through the heavens.

Such falls have happened in cloudy as well as in clear weather, which leads to the belief that they are wholly unconnected with the state of the atmosphere.

The most remarkable circumstance respecting them is, that they invariably resemble each other in certain marked respects, of external properties and chemical composition, so as to render it possible for a mineralogist or a chemist to recognise them with certainty, even though he has not been informed of their origin and fall. They invariably contain, disseminated throughout their substance, an alloy of iron and nickel, which, it is believed, has never yet been discovered among the productions of our earth. This alloy varies in the proportion it bears to the earthy matters, being sometimes scarcely to be recognised without the aid of a microscope, and at other times forming more than one-half of the stone. They are generally coated on the outside with a thin black incrustation; and indentations may often be noticed, as though the mass had been impressed with the fingers.

These constant characters, as respects both their fall and their chemical and mechanical composition, indicate a common origin, and have given rise to numberless theories to account for their phenomena. Some have attributed them to terrestrial, and others to lunar volcanoes. But if our earth's volcanoes had anything to do with them, we could not have avoided noticing the coincidence of the one with the other; and there is, besides, a want of similarity between meteoric stones and ordinary volcanic exuviæ. And as to their coming from lunar volcanoes, the theory might be tenable if we had only to account for the showers of stones which come to our earth's surface, but these are, probably, only a very small part of the masses from which they descend; and since we experience a shower of these stones every few months in some part of the world, it is obvious that, at this rate, the whole mass of the moon would soon be shot away.

With regard to the theory of these bodies having formed in our own atmosphere, there is one principal objection, viz., that the velocity with which they strike the earth, as shown by the depth to which they have sometimes penetrated, indicates their having fallen from heights far exceeding the limits of the terrestrial atmosphere.

The remaining theory is the one now generally accepted, viz., that they are terrestrial comets, or detached portions of solar comets. And the periodicity of their occurrence favours this idea. The second week in August, and the beginning of November, are the periods in each year when these meteoric falls seem to be most frequent, plainly indicating by their regular recurrence, that our earth is then passing near the line of the annual orbit of a large company of these comets or miniature planets.

They all appear, on examination, to have undergone the action of fire, through violent collision with our atmosphere.

It is recorded that a meteoric mass fell at Agram in Hungary in 1751, weighing 71 pounds. But the largest and heaviest ever known fell in Mexico, and was computed to weigh about 15 tons!

Our Earth's Internal Heat.—" The Monkwearmouth colliery reaches a depth of 1800 feet below the surface of the ground, and nearly as much below the level of the sea."

" The observed temperature of the strata at this depth

agrees closely with what has been ascertained at other places, showing that the increase of heat takes place at the rate of about 10° (Fahr.) for every 600 feet downwards vertically from the surface. Assuming, then, the temperature of subterranean fusion to be equal to 3000° (Fahr.) and that the heat goes on increasing uniformly (which, however, is not quite certain), then the thickness of the *film* separating us from the fiery ocean beneath, will be about 35 miles,—a thickness which may fairly be represented by the *skin of a peach*, as compared with the body of the fruit which it encloses." (!)

The above statement was addressed, in 1863, to the British Association, by their President, Sir W. Armstrong.

But this theory of increasing heat at greater depths below the earth's surface is altogether scouted by Sir John Leslie (a Scotchman of scientific eminence), in a work published as late as 1860. Sir John accounts for the increase of heat in deep coal-pits solely by the number of lamps used by the miners, and by the difficulty of introducing fresh cool air from the surface. He even goes so far as to assert that the contrary is the case—maintaining that if one of these very deep shafts were left open and not used for a considerable time, and then examined, greater cold would be found at the bottom of the shaft, than the mean annual temperature of the air at the surface of the ground above. (!)

It was also urged by the supporters of Sir John Leslie's theory, that the running water at the bottom of deep mines was always colder than the air at the same place. This would have been a powerful argument indeed, if it could have been shown that the water at the bottom of a deep coal-pit found its way thither from a still lower point, welling up from the bowels of the earth. As, however, there can be no doubt that the water had flowed *downwards*, from points nearer the surface of the ground, its colder temperature (if it is to prove anything) rather militates against Sir John Leslie's idea than in favour of it. But who shall decide when these two doctors disagree ?

Relation of Climate to Organic Development.—(From Scoffern and Lowe's Practical Meteorology.)—"Even in the narrow region of European travel the intelligent observer will not fail to see distinctive physiognomies. Passing from the cold green sward and modest vegetation of our own isles to the Mediterranean shore, a striking change in the aspect of nature meets the eye. The sturdy oaks and elms of our own forests disappear; the absence of smaller grasses removes the green carpet of our meadows; tall graminaceæ spring up; the aloe and the prickly pear bespeak a mixed condition of heat and drought; and the date palm, barely acclimatised there, gives some faint notion of what the characteristics of a tropical forest must be. It would be an enterprise worthy of a great artist, says Humboldt, to study the aspect and character of all these vegetable groups, not merely in hot-houses or in the descriptions of botanists, but in their native grandeur in the tropical zone. How interesting and instructive to the landscape painter would be a work which should present to the eye, first, separately, and then in combination and contrast, their leading forms! How picturesque is the aspect of tree-ferns spreading their delicate fronds above the laurel-oaks of Mexico, or the groups of plaintains overshadowed by arborescent grasses! It is the artist's privilege, having studied these groups, to analyse them; and thus, in his hands, the grand and beautiful forms of nature resolve themselves into a few simple elements. But when the meteorologist has exhausted his knowledge in laying out climatic groups;—when he has placed in correlation conditions identical in every respect, as he may think—the growth of vegetable forms demonstrates his inability to comprehend many hidden secrets in nature which their delicate organisation reveals. Thus, European olive trees grow luxuriantly at Quito, but they bear neither fruit nor flowers; and a similar remark applies to walnut trees and the hazel in the island of Mauritius. In India the bamboo flowers luxuriantly; but in South America,—where it grows equally well, so far as general vigour can be judged by its appearance,—so rare an event is the flowering of the bamboo, that, during a four years' residence in South America, Humboldt was only able to obtain blossoms once. But, perhaps, the most remarkable example of luxuriant growth, without inflorescence, is furnished by the sugar-cane. The West Indies have come to be considered the region, *par excellence*, of the sugar-cane; yet it seldom bears flowers there, nor, indeed, does it on any part of the American continent,—thus affording a strong presumptive argument in favour of the theory which asserts that no variety of the sugar-cane is indigenous to the New World." In all these so-called freaks of nature, there are subtle influences at work which baffle our most curious research, for the present time at least, and must for ever commend to our admiration the wisdom and power of Him "which doeth great things past finding out; yea, and wonders without number."

The Atomic Theory.—Heat, Light, Electricity, &c.—Much as we have yet to learn respecting these agencies, we know enough already to infer that they cannot be transmitted from the sun to the earth, except by communication from particle to particle. Not that the term "particles" need be understood in the sense of the atomist. Whatever our views as to the nature of particles, we must conceive them to be centres invested with surrounding forces. But we have no reason to believe that these centres are occupied by solid cores of indivisible, incompressible matter, essentially distinct from force.

And why encumber our conceptions of material forces with this needless imagining of a central molecule? If we retain the forces and reject the molecule, we have still every property that we can recognise in matter, by the use of our senses or by the aid of our reason. Viewed in this light, matter is not only a thing *subject* to force, but is itself composed and constituted of force.

The Dynamical theory of heat is, perhaps, the most important discovery of the present century. We now know that each Fahrenheit degree of temperature, in a pound of water, is equivalent to a weight of 772 pounds lifted one foot high, and that these amounts of heat and power are reciprocally convertible into one another. This discovery may be assigned to the present century, because it is only of late that it has been indisputably settled. But if we search back for an earlier conception of the identity of heat and motion, we shall find that similar ideas have been held before, though only in a clouded and undemonstrated form. In the writings of Lord Bacon, we find it stated that heat is to be regarded as motion, and nothing else. In dilating upon this subject, that extraordinary man shows that he had grasped the true theory of heat, to the fullest extent that was compatible with the state of knowledge existing in his day. Even Aristotle (B.C. 350) seems to have entertained the idea, that motion was to be considered as the foundation and cause, not only of heat, but of all other manifestations in matter. And for aught we know, still earlier thinkers may have held the same views. In this age of practical experiment, superadded to close reasoning, we have proved that many of their *guesses* on the subject of heat, light, motion, &c., were actually flashes of truth.

Imponderable Agents or Forces.—These expressions are now

applied generally to indicate the causes, whatever they may be, which give rise to the phenomena of heat, light, electricity, and magnetism; and we must now include actinism, or the radiant influence of the sun, to the operation of which photographic pictures are due, and which is believed to be the chemical agency of a portion of the sun's rays.

The imponderable agents have always presented a field of great interest to the student; and more especially is it so at this present time, seeing that it is the tendency of modern philosophy to refer all these agencies or forces to various modifications of one grand cause.

The correlations between light and heat, electricity and magnetism, are so intimate and so well marked, that it is impossible to avoid the conclusion that they must all be due to a modification of one common agent. We may say that this grand cause is motion; and it may be that we shall never reach a higher step in our search after nature's "causa causarum" than that we have now attained. Must we, then, at one step, pass thence from the regions of the material to those of the spiritual—from matter to mind and will? At our present stage of knowledge we show the humility of true wisdom by allowing this question to be answered affirmatively.

Theories respecting Light.—There are two aspects under which light is now regarded—either as being an emission of material *particles* from the luminous body, or as *undulations* of a very subtle medium (ether) which fills all space, and communicates vibrations from the solar disc to the earth, or other planetary bodies, by an uninterrupted series of waves. Whichever theory is adopted, whether of particles or waves, certain general truths will remain the same. Every particle or wave of light is emitted in straight lines; and from every point of the luminous body these particles or waves are thrown off in all directions. Under both theories, light is subject to two great laws, viz., the law of *Refraction*, and the law of *Reflection*. A ray of light is said to be refracted, when its course is diverted from a straight line by entering some denser medium; and it is said to be reflected, when it is thrown back by any body on which it falls. In order that a ray may be refracted, it must enter a more or less transparent body, such as air, water, or glass; and to be reflected, it must strike upon bodies through which it cannot penetrate, such as bright metal, when, instead of being transmitted, it is thrown back in whole or in part. The beauty of the visible creation is almost entirely the result of refraction and reflection. The refraction of the atmosphere diffuses the solar light throughout the sky, renders the transition from day to night, and from night to day, gradual and pleasing, and gives infinite variety to every hue and tint. In like manner reflection, by its more or less perfect character, gives tone and beauty to what would otherwise be a dazzling multiplicity of images, or a monotonous sameness in the diffusion of light and colour.

Colour of the Atmosphere.—On ascending to the top of a very high mountain, we find that the sky, even in bright daylight, has lost very much of the blue tint which we generally observe when we gaze upwards from a lower level. From this it appears, that were there no atmosphere surrounding our planet, the heavens would be intensely black at all times, whilst the heavenly bodies would be so many brilliant points set in the deepest shadow. The air, however, diffuses and reflects the sun's rays of light, so that a portion of them is received from every part of the sky, causing a general illumination of the entire vault of heaven. This reflected light ought, as we should at first suppose, to be always of a whitish character, like the light of the sun made very much fainter. But we find that the sky, when very clear, has a strong blue tint, which deepens into black the higher we ascend a lofty mountain side. This blueness arises from the peculiar action of the atmosphere upon light. If any solar beam pass through a great thickness of air, the red will be transmitted the most readily, and lost to our vision, whilst the blue is resisted and reflected; and from this cause it is that the reflected rays of the solar light have a bluish tinge instead of being pure white. The red tint that we sometimes see in the sky is caused by the passage of the sun's rays through a more than usually dense medium of air. This appearance, when seen at mid-day, betokens much moisture in the air, and is generally the precursor of wind and rain. But the same ruddy illumination at sunset or sunrise is caused by the multifold thickness of atmospheric air through which the rays are obliquely transmitted along the surface of the earth.

Cause of Twilight.—This is the effect of refraction, the air causing a portion of the sun's light to reach the earth by an angular course, when the sun is below the observer's horizon. It continues until the sun has descended 18°, or for about two hours after sunset in this country.

Cause of the Rainbow.—This beautiful collection of colours is owing to a complicated reflection and decomposition of the sun's ray in passing through the drops of rain. It appears when the sun is unclouded, and rain is falling in the opposite quarter of the heavens, and requires that the observer should be placed between the sun and the shower to be acted upon. When a ray of light enters a drop of rain, it is immediately decomposed into the prismatic colours; and on reaching the other side of the drop, they are partly transmitted and partly reflected, and thrown out at the same side as that at which they entered, when they emerge in the coloured state, and are visible, In the brilliant or *primary rainbow*, the violet strip is the lowest

D

down, or on the inner side of the illuminated arch, and the red is on its outermost edge. When a secondary rainbow appears inclosing the primary one, it is the result of *double* reflection, and therefore is less brilliant. In a very stormy sea, with the sun shining on the broken waters, showers of spray are sometimes tossed upwards by the action of the wind, when the rays of the sun produce *inverted* rainbows, of which more than twenty have often been seen at the same moment. They cannot, however, be compared, for brilliancy, with the ordinary rainbow, being very faint, and but rarely consisting of more than two colours, viz., yellow on the side towards the sea, and pale green on the other side.

"*Actinism,*" *and the* "*Actinic Rays.*"—These expressions are very much in use now-a-days; but though photographers and medical men may attach to them a correct and definite meaning, it is probable that many persons have only a vague idea of what it is they wish to express when they use these terms. A few sentences may suffice to give an outline of the Actinic theory.

It is well known that the solar light is made up of several coloured rays, so blended as to appear white or colourless. Thus we speak of *seven primary colours*, meaning that by the aid of a prism we can decompose the sun's light, so as to produce seven differently coloured rays. It appears, however, that it would be more correct to speak of only *three primary colours,* viz., red, yellow, and blue; and to treat the others, viz., violet, indigo, green, and orange, as compound colours, that is to say, as the result of the intermingling of one or other pair of the three primary colours. *Red, yellow, and blue,* then, are the *three primary colours.* Each of these is proved to have a certain province, and to produce certain effects, distinct from the other two; thus:

The red is the heating ray:

The yellow is the luminous or visual ray :
The blue is the chemical or actinic ray.

These several effects will be more or less intensified or blended together, according to the degree in which any one ray prevails over the others.

We may sum up the chief points regarding the chemical or actinic influence of solar light as follows :—

I. "The rays having different illuminating and calorific powers, they exhibit different degrees and kinds of chemical action."

II. "The most luminous rays exhibit the least chemical action on all *inorganic* matter. The least luminous and non-luminous rays manifest very powerful chemical action on the same inorganic substances."

III. "The most luminous rays influence all substances having an *organic* origin, particularly exciting *vital* power.

IV. "Thus, under modifications, chemical power is traced to every part of the prismatic spectrum; but in some cases this action is positive,—*exciting;* and in others negative, —*depressing*.

V. "The most luminous rays are proved to prevent all chemical change upon inorganic bodies, though exposed at the same time to the influence of the chemical rays."

VI. "Hence actinism,—though regarded as only a phenomenon different to light,—really stands in direct antagonism to light."

VII. "Heat radiations produce chemical change, in virtue of some *combined* action, not yet understood."

VIII. "Actinism (*i.e.*, the blue ray and its shades) is necessary for the healthful germination of the seed. Light (the yellow ray and its compounds) is required to excite the plant to decompose carbonic acid. Caloric (the red ray and its shades) is necessary for developing and carrying out the reproductive functions of the plant."

IX. "Phosphorescence is due to actinism, and not to light."

X. "Electrical phenomena are quickened by actinism, and retarded by light."

The foregoing summary of actinism is taken from a treatise on photography by Professor Hunt.

CONCLUSION.

Use of Meteorology.—It would be quite a mistake to imagine, that the study of meteorology must necessarily be a mere source of amusement for the rich and the idle. We have seen how a knowledge of the hygrometer (dry and wet bulbs) may be turned to good account, both in hot-houses and in sick rooms; and that the hygrometric condition of the atmosphere is really a far surer guide than even the barometer, in judging as to the near approach of rain. And so of the rain-gauge,—it is a valuable assistant in the management of out-door plants and crops, as well as in the construction of tanks and reservoirs, for the regular supply of water on a large scale. The observer may see how beneficial is even a passing shower to growing plants, when he remembers that one-tenth of an inch of rain, in his gauge, represents a deposit of about forty hogsheads per acre.

Or, to take the solar maximum and terrestrial minimum thermometers,—the one will show the necessity of screening delicate ripening fruits, in the hot-house, from the dry glare of the mid-day sun; the other will forewarn the observer, to guard his tender plants from the chilling effect of heavy dew and hoar frost.

All these instruments just named, differ from the barometer in the mode of their indications, in that they warn us of the *present hour;* whereas, the barometer gives us notice of general changes, which may require a week or more for their full development. We might, therefore, call the barometer the seaman's weather-gauge, rather than the landsman's; the more so, as its movements are indicative *not* of moisture but of *pressure;* that is, not of rain, but of wind.

Calendar of Nature.—" Every meteorologist (Mr Lowe says) should endeavour, as much as possible, to record the arrival and departure of migratory birds,—the dates on which trees open and lose their leaves,—the first appearance of such insects as caterpillar-blight," &c. In regard to trees, however, it may be necessary to remind the observer that he should take one *individual* of a species, as the subject of his observations, say a Beech or a Lime, and not indifferently one or another in successive years. For different individuals of the *same* species will come into leaf at considerable intervals. Thus, one beech will precede its neighbour beech in the opening of its leaves by ten or even fifteen days, though growing side by side.

This calendar of nature should be entered into the meteorological book, in such a way that one page may serve as a record for many years. It might be headed " Spring page," and " Autumn page," and occur, the one between May and June, and the other between November and December.

" Every observer who, for a series of years, pursues such a course of observation, will find that he has done some good to his kind. And if any method could be devised for bringing together such a body of observations throughout the kingdom, they would form a valuable mass of facts for the naturalist and meteorologist to generalise upon." Such a method has been contrived by Mr Symons, and is being carried out by him, with the aid of a thousand observers, distributed throughout the three kingdoms.

Law implies Mind (From the " Bridgewater Treatises").—Though at present comparatively in an infant state, the science of meteorology is no doubt as capable of being accurately reduced to certain great laws, as the science of optics, acoustics, or astronomy, if only our observations could be as continuous and exact as the observations which have led to the construction of the other sciences. And the difficulties to be overcome in finding laws which shall be of universal application, over so wide a field as that embraced in meteorology, should rather tend to increase our admiration at the infinite wisdom which could provide a system so grandly complex, and yet so all-sufficient for the minutest details. Now this connection between the laws of the material world and an Intelligence which preconceived and instituted the laws, is not only the belief of the untutored mass of mankind, but has struck no less forcibly those who have studied nature with the more systematic attention, and with the peculiar views which belong to science. The laws which such persons study seem, indeed, most naturally to lead to the conviction of a Supreme Intelligence, which originally gave each law its form. What we call a law is, in truth, a form of expression, including a number of facts of a like kind. The facts are separate; but the unity of view by which we associate them—the character of generality and of law—resides in those relations which are the object of the intellect. The law when once apprehended by us, takes in our mind the place of the facts themselves, and is said to *govern* or determine them, because it determines our anticipations of what they will be. But we cannot, it would seem, receive a law founded on such intelligible relations, as being itself able to govern and determine the facts themselves any otherwise than by supposing also an intelligence, by which these relations are contemplated, and these consequences realised and provided for. We cannot then represent to ourselves the universe as governed by general laws, otherwise than by supposing an intelligent and conscious Deity, by whom these laws were originally contemplated, established, and applied. This perhaps will appear more clear, when it is considered that in every branch of physical science, be it astronomy or meteorology, or the like, the laws of which we speak are often of an abstruse and complex kind, depending upon relations of time, space, number, and other properties. These relations are often combined so variously and so curiously, that the most subtle reasonings and calculations which we can form are requisite in order to trace their results. Can such laws be conceived to be instituted without any exercise of knowledge and intelligence? Can material objects apply geometry and calculation to themselves? Can the lenses of the eye, for instance, be formed and adjusted with an exact suitableness to their refractive powers, while there is in the agency which has formed them no consciousness of the laws of light, of the course of rays, of the visible properties of things? This appears to be altogether inconceivable. Every particle of matter possesses an almost endless train of properties, each acting according to its peculiar and fixed laws. For every atom of the same kind of matter these laws are invariable and perpetually the same. This constant and precise resemblance—this variation, equally constant and equally regular—suggests irresistibly the conception of some cause, independent of the atoms themselves, by which their similarity and dissimilarity, the agreement and difference of their deportment, under the same circumstances, have been determined. Such a view of the constitution of matter (as was remarked by Herschel in his "Study of Natural Philosophy") effectually destroys the idea of its eternal and self-existent nature, " by giving to each of its atoms the essential character at once of a *manufactured article* and a *subordinate agent.*" Clarke, the friend and disciple of Newton, most strenuously urges the same truth, that "all things which we commonly say are the effects of the natural powers of matter, and laws of motion, are indeed (if we will speak strictly and properly) the effects of God's acting upon matter continually and at every moment, either immediately by Himself, or mediately by some created intelligent being. Consequently, there is no such thing as the cause of nature or the power of nature," independent, that is, of the effects produced by the will and providence of God. And Dugald Stewart has adopted and illustrated the same opinion, quoting with admiration the well-known passage of Pope concerning the divine agency, which

> " Lives through all life, extends through all extent,
> Spreads undivided, operates unspent."

METEOROLOGICAL TABLES.

TABLE I.—PREVAILING WINDS IN STRATHEARN.

If we divide the Winds into Three Classes, and call them "Moist Westerlies," "Dry Westerlies," and "Cold Easterlies," the accompanying Table will show the Average Number of Days belonging to each Class, in the several Months, during Six Years (1860-1865).

Mean of Six Years.

Muthill, near Crieff, Perthshire, 245 feet above sea-level.	Moist Westerlies (including) S., S.S.W., S.W., and W.S.W.	Dry Westerlies (including) W., W.N.W., N.W., N.N.W., and N.	Cold Easterlies (including) N.N.E., N.E., E.N.E., E., E.S.E., S.E., and S S.E.
January,	9·1	9·8	12·0
February,	6·8	9·0	12·5
March,	6·8	11·3	12·8
April,	8·1	8·7	13·1
May,	12·0	6·1	12·8
June,	9·8	9·1	11·0
July,	10·3	12·5	8·1
August,	9·8	12·7	8·5
September,	11·8	11·2	7·0
October,	8·0	10·5	12·5
November,	7·0	9·5	13·5
December,	9·7	10·3	11·0
Mean total days in each class, during six years,	109	121	135

(STRATHEARN.)—MUTHILL RAINFALL, IN INCHES.

Mean Monthly Fall of Six Years (1860–1865).

Jan.	Feb.	March	April.	May.	June.	July.	Aug.	Sept.	Oct.	Nov.	Dec.
5·59	3·83	2·65	2·07	2·20	3·17	2·77	4·08	3·69	4·43	3·95	5·26
Mean Annual Total (43·69) Fall, in Inches.											

TABLE II.—*Six Years, and Mean of Barometer, "Corrected and Reduced."*

Muthill, Perthshire, 245 feet above Sea-level.	1860. Extremes.	1860. Mean.	1861. Extremes.	1861. Mean.	1862. Extremes.	1862. Mean.	1863. Extremes.	1863. Mean.	1864. Extremes.	1864. Mean.	1865. Extremes.	1865. Mean.	6 Years' Mean. Average Extremes.	6 Years' Mean. Mean.
January	30·34 / 28·41	29·54	30·45 / 29·41	30·17	30·46 / 28·87	29·74	30·45 / 28·60	29·51	30·52 / 29·18	29·92	29·97 / 28·20	29·35	30·36 / 28·78	29·70
February	30·83 / 28·61	30·07	30·46 / 28·79	29·73	30·71 / 29·26	30·18	30·64 / 29·19	30·14	30·43 / 28·63	29·80	30·49 / 28·76	29·74	30·59 / 28·87	29·94
March	30·56 / 28·62	29·64	30·27 / 28·70	29·50	30·43 / 29·01	29·74	30·35 / 29·09	29·82	30·02 / 28·91	29·54	30·16 / 29·06	29·81	30·29 / 28·90	29·67
April	30·54 / 28·61	30·04	30·57 / 29·82	30·23	30·42 / 29·20	29·93	30·22 / 29·16	29·85	30·24 / 29·22	29·93	30·33 / 29·67	30·02	30·38 / 29·28	30·00
May	30·47 / 29·19	29·82	30·32 / 29·43	30·12	30·24 / 29·38	29·83	30·42 / 29·27	30·02	30·22 / 29·56	29·90	30·18 / 29·32	29·73	30·30 / 29·36	29·90
June	30·22 / 29·04	29·64	30·21 / 29·56	29·96	30·13 / 29·04	29·77	30·14 / 29·37	29·77	30·05 / 29·33	29·73	30·43 / 29·62	30·08	30·19 / 29·33	29·82
July	30·39 / 29·56	30·01	30·00 / 29·17	29·60	30·15 / 29·42	29·74	30·44 / 29·65	30·11	30·23 / 29·23	29·85	30·23 / 29·35	29·75	30·24 / 29·40	29·84
August	30·03 / 28·74	29·53	30·11 / 29·35	29·77	30·28 / 29·36	29·92	30·16 / 29·34	29·79	30·45 / 29·44	29·87	30·08 / 29·31	29·70	30·18 / 29·26	29·76
September	30·32 / 29·22	29·87	30·13 / 28·94	29·69	30·51 / 29·46	30·04	30·24 / 28·73	29·57	30·19 / 29·05	29·66	30·38 / 29·47	30·04	30·29 / 29·15	29·81
October	30·36 / 28·92	29·83	30·40 / 29·08	29·99	30·46 / 28·57	29·66	30·21 / 28·53	29·67	30·43 / 28·57	29·84	30·17 / 28·64	29·53	30·34 / 28·72	29·75
November	30·62 / 28·98	29·99	30·38 / 28·95	29·54	30·49 / 28·87	30·03	30·43 / 23·96	29·89	30·62 / 28·54	29·58	30·37 / 28·48	29·67	30·48 / 28·79	29·78
December	30·34 / 28·78	29·74	30·71 / 29·03	30·16	30·35 / 28·94	29·85	30·37 / 28·85	29·89	30·48 / 29·21	29·85	30·64 / 28·70	29·92	30·48 / 28·92	29·90
Each year's Extremes, and Mean	30·83 / 28·41	29·81	30·71 / 28·70	29·87	30·71 / 28·57	29·87	30·64 / 28·53	29·84	30·62 / 28·54	29·79	30·64 / 28·20	29·78	30·83 / 28·20	29·82

METEOROLOGICAL TABLES.

TABLE III.—*Seven Years, and Mean of Thermometer (compared with Highfield House Observatory—Notts).*

Muthill, Perthshire, 245 feet above Sea-level.	1859. Ex-tremes.	1859. Mean.*	1860. Ex-tremes.	1860. Mean.	1861. Ex-tremes.	1861. Mean.	1862. Ex-tremes.	1862. Mean.	1863. Ex-tremes.	1863. Mean.	1864. Ex-tremes.	1864. Mean.	1865. Ex-tremes.	1865. Mean.	Strathearn. 7 Years' Ex-tremes.	Strathearn. 7 Years' Mean.	England. 21 Years' Ex-tremes.	England. 55 Years' Mean.
January	52/27	39	48/14	32·8	49/+4	33·5	46/26	35·8	48/22	35·8	45/11	31·8	49/10	34·6	52/+4	34·8	56·1/−4·0	36·7
February	54/21	39·3	46/−12	30·5	48/19	36·0	52/19	38·5	52/20	39·0	46/13	29·8	47/+5	32·5	54/−12	35·1	61·0/+6·0	38·6
March	58/24	43	54/22	36·5	50/27	39·0	50/17	34·7	56/24	41·7	48/10	33·9	52/19	37·3	58/10	38·1	71·5/13·0	42·1
April	63/28	40·5	60/22	39·2	62/25	45·3	71/26	43·7	58/29	43·0	65/27	44·5	71/21	47·1	71/21	43·4	79·0/20·4	47·1
May	74/33	55	70/34	50	73/26	49·5	69/32	49·8	66/31	48·7	74/29	49·7	73/25	51·5	74/25	50·6	89·3/28·8	53·1
June	74/43	58·2	72/42	52·5	81/47	58·5	72/41	52·5	74/39	54·8	66/30	52·5	80/32	58·7	81/32	55·4	92·2/30·5	58·6
July	76/45	60	78/46	58·8	75/42	57·2	72/40	54·0	82/32	58·5	79/41	55·8	76/31	58·6	82/31	57·6	91·0/36·3	61·1
August	73/42	56·5	71/39	55·7	76/46	58·0	74/40	56·8	72/35	56·5	73/34	52·5	69/29	55·5	76/29	55·9	92·5/33·0	60·2
September	72/35	51	75/29	49·3	73/36	54·7	74/33	52·2	65/33	49·5	64/36	49·8	70/31	56·6	75/29	51·9	85·0/32·0	56·4
October	61/20	41	68/28	44·5	64/25	47·8	67/28	44·7	56/24	43·8	58/22	44·4	62/19	44·5	68/19	44·4	77·5/19·4	49·5
November	51/19	36·5	48/18	36·5	50/19	36·5	51/14	31·5	54/18	40·0	53/17	33·7	54/14	39·2	54/14	37·0	61·2/13·2	42·2
December	50/+8	32	48/+2	30·2	53/16	33·3	51/30	40·0	51/21	37·7	53/18	37·7	53/18	42·0	53/+2	36·1	60·2/−8·0	39·2
Each year's Extremes, and Mean	76/+8	46·0	78/−12	43·04	81/+4	45·78	74/14	44·52	82/18	45·75	79/10	43·42	80/+5	46·5	82/−12	45·0	92·5/−8·0	48·73

* The Extremes are those of the Shaded Maximum and *Grass* Minimum; but the Means are those of the Shaded Maximum and *Shaded* Minimum.

The following Hygrometrical Tables are not intended so much to help in finding the Dew-point, as to give the amateur observer an *approximate* idea of the degree to which the air is "saturated with moisture," taking 100 to represent complete saturation.

For finding the dew-point, the observer had better take the small amount of trouble required by the use of the "Table of Factors," given on an earlier page.

But when the observer's dry and wet bulb thermometers give, as their readings, some intermediate point between the readings supplied in these columns, he will readily find the amount of saturation by calculating the proportional increase or decrease, in accordance with the column headed "Less Saturation, for an increase of 1° in dry."

TABLE IV.—HYGROMETER DEW-POINT AND DEGREE OF SATURATION.

(Abridged from Mr Glaisher's Tables.)

Readings of Thermometers.		Temperature of Dew-point.	Degree of Saturation.	Less Saturation, for increase of 1°, in Dry.	Readings of Thermometers.		Temperature of Dew-point.	Degree of Saturation.	Less Saturation, for increase of 1°, in Dry.
Dry.	Wet.				Dry.	Wet.			
15°	15·0°	15·0°	100	−33	18°	17·5°	13·7°	83	27
	14·5	10·6	81	26		17·0	9·5	68	21
	14·0	6·2	67	22		16·5	5·2	56	16
	13·5	1·8	54	16		16·0	1·0	46	11
16	16·0	16·0	100	−32	19	19·0	19·0	100	−32
	15·5	11·7	82	26		18·5	14·9	83	26
	15·0	7·3	67	22		18·0	10·7	68	20
	14·5	3·0	56	19		17·5	6·4	56	17
17	17·0	17·0	100	−32		17·0	2·3	47	13
	16·5	12·7	83	27	20	20·0	20·0	100	−30
	16·0	8·4	68	22		19·5	15·9	83	25
	15·5	4·0	55	17		19·0	11·9	68	20
18	18·0	18·0	100	−32		18·5	7·8	57	16

METEOROLOGICAL TABLES.

Readings of Thermometers.		Temperature of Dew-point.	Degree of Saturation.	Less Saturation, for Increase of 1°, in Dry.	Readings of Thermometers.		Temperature of Dew-point.	Degree of Saturation.	Less Saturation, for Increase of 1°, in Dry.
Dry.	Wet.				Dry.	Wet.			
20	18·0	3·7	48	13	27	26·0	21·4	79	15
	17·5	0·0	40	10		25·5	18·6	70	13
21	21·0	21·0	100	−29		25·0	15·8	61	11
	20·5	17·1	84	24		24·5	13·0	54	9
	20·0	13·1	70	20	28	28·0	28·0	100	−18
	19·5	9·2	58	16		27·5	25·5	89	14
	19·0	5·2	49	13		27·0	22·9	80	13
	18·5	1·3	42	11		26·5	20·3	72	11
22	22·0	22·0	100	−28		26·0	17·8	64	10
	21·5	18·2	85	24		25·5	15·2	57	8
	21·0	14·4	71	19	29	29·0	29·0	100	−17
	20·5	10·6	60	16		28·5	26·7	91	14
	20·0	6·8	50	14		28·0	24·4	82	13
	19·5	3·0	40	12		27·5	22·1	75	12
23	23·0	23·0	100	−27		27·0	19·7	67	9
	22·5	19·3	85	23		26·5	17·5	60	8
	22·0	15·7	72	19		26·0	14·0	54	6
	21·5	12·1	61	15	30	30·0	30·0	100	−15
	21·0	8·4	52	13		29·5	27·9	91	13
	20·5	4·8	44	11		29·0	25·9	83	11
24	24·0	24·0	100	−26		28·5	23·8	76	10
	23·5	20·5	85	22		28·0	21·7	69	7
	23·0	17·1	73	18		27·5	19.6	63	5
	22·5	13·6	63	15	31	31·0	31·0	100	−13
	22·0	10·2	53	12		30·5	29·1	92	12
	21·5	6·7	46	11		30·0	27·3	85	9
	21·0	3·2	39	9		29·5	25·5	78	8
25	25·0	25·0	100	−24		29·0	23·6	72	7
	24·5	21·7	86	20		28·5	21·7	67	5
	24·0	18·5	74	17	32	32·0	32 0	100	−11
	23·5	15·2	64	14		31·5	30·3	93	10
	23·0	11·9	55	11		31·0	28·7	87	8
	22·5	8·7	48	9		30·5	27·0	80	7
26	26·0	26·0	100	−21		30·0	25·4	75	6
	25·5	23·0	88	17		29·5	23·7	70	5
	25·0	19·9	76	15	33	33·0	33·0	100	−11
	24·5	16·9	66	12		32·5	31·5	94	10
	24 0	13·8	58	11		32·0	30·0	89	9
	23·5	10·8	51	10		31·5	28·5	83	8
27	27·0	27·0	100	−20		31·0	27·0	78	7
	26·5	24·2	89	17		30·5	25·5	73	6

86　　　　　　METEOROLOGICAL TABLES.

Readings of Thermometers.		Temperature of Dew-point.	Degree of Saturation.	Less Saturation, for increase of 1°, in Dry.	Readings of Thermometers.		Temperature of Dew-point.	Degree of Saturation.	Less Saturation, for increase of 1°, in Dry.
Dry.	Wet.				Dry.	Wet.			
34°	34·0°	34·0°	100	−10	40°	30°	17·1°	38	3
	33·5	32·6	94	9		28	12·5	31	2
	33·0	31·2	89	9	41	41	41·0	100	−8
	32·5	29·8	84	9		39	36·5	84	6
	32·0	28·5	79	7		37	32·0	70	5
	31·5	27·1	75	6		35	27·4	58	4
	31·0	25·5	72	6		33	22·9	48	3
35	35·0	35·0	100	−10		31	18·4	39	3
	34·5	33·7	95	9		29	13·9	31	2
	34·0	32 4	90	8	42	42	42·0	100	−8
	33·5	31·1	85	7		40	37·5	85	7
	33·0	29·8	80	7		38	33·1	72	6
	32·5	28·5	75	6		36	28·6	60	5
	32·0	27·2	72	6		34	24·2	49	4
36	36	36·0	100	−9		32	19·7	40	3
	34	31·0	82	7		30	15·2	33	2
	32	26·0	66	5		28	10·8	27	2
	30	21·0	53	4	43	43	43·0	100	−8
	28	16·0	42	3		41	38·6	84	7
37	37	37·0	100	−9		39	34·2	71	6
	35	32·2	83	7		37	29·8	59	5
	33	27·3	68	6		35	25·4	49	4
	31	22·5	55	5		33	21·0	41	4
	29	17·6	44	3		31	16·6	34	3
38	38	38 0	100	−8		29	12·2	28	3
	36	33·3	83	6	44	44	44 0	100	−8
	34	28·6	68	5		42	39·6	84	6
	32	23·8	56	4		40	35·3	71	5
	30	19·1	45	3		38	30·9	59	4
	28	14·4	37	3		36	26·6	49	4
39	39	39·0	100	−8		34	22·2	41	3
	37	34·4	84	8		32	17·8	34	3
	35	29·7	70	7		30	13·5	28	3
	33	25·1	57	6	45	45	45·0	100	−7
	31	20·4	47	5		43	40·7	85	6
	29	15·8	38	4		41	36·4	72	5
40	40	40·0	100	−8		39	32·0	60	4
	38	35·4	84	7		37	27·7	50	4
	36	30·8	69	5		35	23·4	42	3
	34	26·3	57	4		33	19·1	34	3
	32	21·7	46	3		31	14·8	28	3

METEOROLOGICAL TABLES.

Readings of Thermometers.		Temperature of Dew-point.	Degree of Saturation.	Less Saturation, for increase of 1°, in Dry.	Readings of Thermometers.		Temperature of Dew-point.	Degree of Saturation.	Less Saturation, for increase of 1°, in Dry.
Dry.	Wet.				Dry	Wet.			
46°	46°	46·0°	100	−7	51°	49°	46·9°	86	6
	44	41·7	86	7		47	42·8	74	5
	42	37·4	73	6		45	38·8	63	4
	40	33·2	61	5		43	34·7	54	4
	38	28·9	51	4		41	30·6	46	3
	36	24·6	43	3		39	26·5	38	3
	34	20·3	35	3		37	22·4	32	2
	32	16·0	29	3	52	52	52·0	100	−7
47	47	47·0	100	−7		50	48·0	86	6
	45	42·8	86	6		48	43·9	74	5
	43	38·5	73	5		46	39·9	64	4
	41	34·3	61	5		44	35·9	54	4
	39	30·0	51	4		42	31·8	46	3
	37	25·8	43	3		40	27·8	39	3
	35	21·6	36	3		38	23·7	33	2
	33	17·3	30	2	53	53	53·0	100	−7
48	48	48·0	100	−7		51	49·0	86	6
	46	43·8	86	7		49	45·0	74	5
	44	39·6	73	6		47	41·0	64	5
	42	35·4	62	5		45	37·0	55	4
	40	31·2	52	4		43	33·0	47	3
	38	27·0	44	3		41	29·0	39	3
	36	22·8	36	2		39	25·0	33	2
	34	18·6	30	2	54	54	54·0	100	−7
49	49	49·0	100	−7		52	50·0	86	6
	47	44·8	86	6		50	46·1	74	5
	45	40·7	73	5		48	42·1	64	4
	43	36·5	62	4		46	38·2	55	3
	41	32·4	53	4		44	34·2	47	3
	39	28·2	45	3		42	30·2	40	3
	37	24·0	37	3		40	26·3	34	2
	35	19·9	31	2	55	55	55·0	100	−7
50	50	50·0	100	−7		53	51·1	87	5
	48	45·9	86	6		51	47·2	75	5
	46	41·8	74	5		49	43·2	65	4
	44	37·6	63	5		47	39·3	56	4
	42	33·5	53	4		45	35·4	48	3
	40	29·4	45	3		43	31·5	41	3
	38	25·3	37	2		41	27·6	35	2
	36	21·2	31	2	56	56	56·0	100	−7
51	51	51·0	100	−7		54	52·1	87	6

METEOROLOGICAL TABLES.

Readings of Thermometers		Temperature of Dew-point.	Degree of Saturation.	Less Saturation, for increase of 1°, in Dry.	Readings of Thermometers.		Temperature of Dew-point.	Degree of Saturation.	Less Saturation, for increase of 1°, in Dry.
Dry.	Wet.				Dry.	Wet.			
56°	52°	48·2°	75	5	61°	55°	49·8°	67	4
	50	44·4	65	4		53	46·0	58	4
	48	40·5	56	4		51	42·3	50	3
	46	36·6	48	3		49	38·6	44	3
	44	32·7	41	3		47	34·8	38	2
	42	28·8	35	2	62	62	62·0	100	− 6
57	57	57·0	100	− 6		60	58·3	88	6
	55	53·2	87	6		58	54·6	77	5
	53	49·3	75	5		56	50·8	67	4
	51	45·5	65	4		54	47·1	58	4
	49	41·6	57	4		52	43·4	50	3
	47	37·8	49	3		50	39·7	44	3
	45	34·0	42	2		48	36·0	38	3
	43	30·1	36	2	63	63	63·0	100	− 6
58	58	58·0	100	− 6		61	59·3	88	5
	56	54·2	87	6		59	55·6	77	4
	54	50·4	76	5		57	51·9	67	4
	52	46·6	66	4		55	48·2	59	4
	50	42·8	57	4		53	44·5	51	3
	48	39·0	49	3		51	40·8	44	2
	46	35·2	43	3		49	37·1	38	2
	44	31·4	37	2	64	64	64·0	100	− 6
59	59	59·0	100	− 6		62	60·3	88	5
	57	55·2	88	6		60	56·7	77	5
	55	51·4	76	5		58	53·0	67	4
	53	47·7	66	4		56	49·4	59	4
	51	43·9	57	4		54	45·7	51	3
	49	40·1	49	3		52	42·0	45	3
	47	36·3	43	3		50	38·4	39	2
	45	32·5	37	2	65	65	65·0	100	− 6
60	60	60·0	100	− 6		63	61·4	88	6
	58	56·2	88	6		61	57·7	78	5
	56	52·5	76	5		59	54·1	68	4
	54	48·7	66	4		57	50·4	59	4
	52	45·0	58	4		55	46·8	51	3
	50	41·2	50	3		53	43·2	45	3
	48	37·4	43	3		51	39·5	39	2
	46	33·7	37	2	66	66	66·0	100	− 6
61	61	61·0	100	− 6		64	62·4	88	5
	59	57·3	88	6		62	58·8	78	5
	57	53·5	77	5		60	55·1	68	4

METEOROLOGICAL TABLES.

Readings of Thermometers.		Temperature of Dew-point.	Degree of Saturation.	Less Saturation, for increase of 1°, in Dry.	Readings of Thermometers.		Temperature of Dew-point.	Degree of Saturation.	Less Saturation, for increase of 1°, in Dry.
Dry.	Wet.				Dry.	Wet.			
66°	58°	51·5°	60	3	71°	61°	53·4°	53	3
	56	47·9	52	3		59	46·9	47	2
	54	44·3	45	3		57	46·4	41	2
	52	40·7	40	2	72	72	72·0	100	−6
67	67	67·0	100	−6		70	68·5	89	5
	65	63·4	88	6		68	65·0	79	4
	63	59·8	78	5		66	61·5	69	4
	61	56·2	68	5		64	58·0	61	4
	59	52·6	60	4		62	54·5	54	3
	57	49·0	52	4		60	51·0	48	3
	55	45·4	46	3		58	47·5	42	2
	53	41·8	40	3	73	73	73·0	100	−6
68	68	68·0	100	−6		71	69·5	89	5
	66	64·4	88	6		69	66·0	79	4
	64	60·8	78	5		67	62·6	70	4
	62	57·3	68	5		65	59·1	62	3
	60	53·7	60	4		63	55·6	54	3
	58	50·1	52	4		61	52·1	48	3
	56	46·5	46	3		59	48·6	42	2
	54	42·9	40	3	74	74	74·0	100	−6
69	69	69·0	100	−6		72	70·5	89	5
	67	65·4	88	5		70	67·1	79	5
	65	61·9	78	5		68	63·6	70	4
	63	58·3	68	4		66	60·2	62	4
	61	54·8	60	4		64	56·7	55	3
	59	51·2	53	3		62	53·2	48	3
	57	47·6	47	3		60	49·8	43	2
	55	44·1	41	3	75	75	75·0	100	−6
70	70	70·0	100	−6		73	71·6	89	5
	68	66·5	88	5		71	68·1	79	4
	66	62·9	78	5		69	64·7	70	4
	64	59·4	69	4		67	61·2	62	3
	62	55·8	61	4		65	57·8	55	3
	60	52·3	53	3		63	54·4	49	3
	58	48·8	47	3		61	50·9	43	2
	56	45·2	41	2	76	76	76·0	100	−5
71	71	71·0	100	−6		74	72·6	89	5
	69	67·5	88	5		72	69·2	79	4
	67	64·0	78	5		70	65·7	71	4
	65	60·4	69	4		68	62·3	63	3
	63	56·9	61	3		66	58·9	55	3

METEOROLOGICAL TABLES.

Readings of Thermometers.		Temperature of Dew-point.	Degree of Saturation.	Less Saturation, for increase of 1°, in Dry.	Readings of Thermometers.		Temperature of Dew-point.	Degree of Saturation.	Less Saturation, for increase of 1°, in Dry.
Dry.	Wet.				Dry.	Wet.			
76°	64°	55°·5	49	3	81°	67°	57°·5	44	2
	62	52·1	43	2	82	82	82·0	100	−5
77	77	77·0	100	−5		80	78·7	90	5
	75	73·6	89	5		78	75·3	80	4
	73	70·2	79	4		76	72·0	72	4
	71	66·8	71	4		74	68·6	64	3
	69	63·4	63	3		72	65·3	57	3
	67	60·0	56	3		70	61·9	51	3
	65	56·6	50	2		68	58·6	45	3
	63	53·2	44	2	83	83	83·0	100	−5
78	78	78·0	100	−5		81	79·7	90	5
	76	74·6	89	5		79	76·3	80	4
	74	71·2	79	4		77	73·0	72	4
	72	67·8	71	4		75	69·7	64	3
	70	64·4	63	3		73	66·3	57	3
	68	61·1	56	3		71	63·0	51	3
	66	57·7	50	3		69	59·7	45	2
	64	54·3	44	3	84	84	84·0	100	−5
79	79	79·0	100	−5		82	80·7	90	5
	77	75·6	90	5		80	77·4	80	4
	75	72·3	80	4		78	74·0	72	4
	73	68·9	71	4		76	70·7	64	3
	71	65·5	63	3		74	67·4	57	3
	69	62·1	56	3		72	64·1	51	3
	67	58·7	50	3		70	60·8	45	2
	65	55·4	44	3	85	85	85·0	100	−5
80	80	80·0	100	−5		83	81·7	90	5
	78	76·6	90	5		81	78·4	80	4
	76	73·3	80	4		79	75·1	72	4
	74	69·9	71	4		77	71·8	64	3
	72	66·5	63	3		75	68·5	58	3
	70	63·2	56	3		73	65·2	52	3
	68	59·8	50	3		71	61·8	46	2
	66	56·4	44	2	86	86	86·0	100	−5
81	81	81·0	100	−5		84	82·7	90	5
	79	77·7	90	5		82	79·4	80	4
	77	74·3	80	4		80	76·1	72	4
	75	70·9	72	4		78	72·8	64	3
	73	67·6	64	3		76	69·5	58	3
	71	64·2	56	3		74	66·2	52	2
	69	60·9	50	3		72	62·9	46	2

METEOROLOGICAL TABLES.

Readings of Thermometers.		Temperature of Dew-point.	Degree of Saturation.	Less Saturation, for increase of 1°, in Dry.	Readings of Thermometers.		Temperature of Dew-point.	Degree of Saturation.	Less Saturation, for increase of 1°, in Dry.
Dry.	Wet.				Dry.	Wet.			
87°	87°	87°·0	100	−5	92°	92°	92°·0	100	−5
	85	83·7	90	5		90	88·8	90	4
	83	80·4	81	4		88	85·5	81	4
	81	77·1	73	4		86	82·3	73	4
	79	73·9	65	4		84	79·1	66	3
	77	70·6	58	3		82	75·9	59	3
	75	67·3	52	3		80	72·6	53	3
	73	64·0	46	2		78	69·4	47	2
88	88	88·0	100	−5	93	93	93·0	100	−5
	86	84·7	90	5		91	89·8	90	4
	84	81·5	81	4		89	86·6	82	4
	82	78·2	73	4		87	83·4	74	4
	80	74·9	65	3		85	80·1	66	3
	78	71·6	58	3		83	76·9	60	3
	76	68·4	52	3		81	73·7	54	3
	74	65·1	46	2		79	70·5	48	2
89	89	89·0	100	−5	94	94	94·0	100	−5
	87	85·7	90	5		92	90·8	90	4
	85	82·5	81	4		90	87·6	82	4
	83	79·2	73	4		88	84·4	74	4
	81	76·0	65	3		86	81·2	66	3
	79	72·7	58	3		84	78·0	60	3
	77	69·4	52	2		82	74·8	54	3
	75	66·2	46	2		80	71·5	48	2
90	90	90·0	100	−5	95	95	95·0	100	−5
	88	86·8	90	4		93	91·8	91	5
	86	83·5	81	4		91	88·6	82	4
	84	80·3	73	4		89	85·4	74	4
	82	77·0	65	3		87	82·2	66	3
	80	73·7	59	3		85	79·0	60	3
	78	70·5	53	3		83	75·8	54	2
	76	67·2	47	2		81	72·6	48	2
91	91	91·0	100	−5	96	96	96·0	100	−5
	89	87·8	90	4		94	92·8	90	4
	87	84·5	82	4		92	89·6	82	4
	85	81·3	74	4		90	86·4	74	3
	83	78·0	66	3		88	83·3	66	3
	81	74·8	59	3		86	80·1	60	3
	79	71·6	53	3		84	76·9	54	3
	77	68·3	47	2		82	73·7	49	2

RAINFALL IN THE BRITISH ISLES.—Average of 14 Years (1850-1863).

ENGLAND.			SCOTLAND.		
Station.	County.	Mean Annual Fall in Inches.	Station.	County.	Mean Annual Fall in Inches.
Enfield,	Middlesex,	23·75	Thurston,	Haddington,	27·55
Chichester,	Sussex,	27·27	Glencorse,	Edinburgh,	37·13
Ventnor,	Hants,	28·97	Inveresk,	Edinburgh,	26·40
Hitchin,	Herts,	24·23	Bothwell Castle,	Lanark,	28·45
Banbury,	Oxon,	25·01	Mansfield, Largs,	Ayr,	45·70
Epping,	Essex,	23·88	Castle Toward,	Argyll,	50·77
Holkham,	Norfolk,	25·63	Torosay Castle,	Argyll,	79·79
Baverstock,	Wilts,	29·42	Pittenweem,	Fife,	24·85
Goodámoor,	Devon,	56·34	Stanley,	Perth,	29·85
Exeter Institution,	Devon,	28·00	Craigton,	Forfar,	32·75
Bodmin,	Cornwall,	44·67	Castle Newe,	Aberdeen,	33·40
Brislington,	Somerset,	29·60	Sandwick,	Orkney,	36·76
Cirencester,	Gloucester,	30·05	Portree,	Isle of Skye,	124·00 (!)
Shiffnal,	Shropshire,	24·44	Bressay,	Shetland,	37·82
Orleton,	Worcester,	29·78	IRELAND.		
Wigston,	Leicester,	26·18			
Derby,	Derby,	24·60	Portlaw,	Waterford,	41·62
Liverpool,	Lancashire,	24·92	Killaloe,	Clare,	41·57
Bolton,	Lancashire,	46·42	Black Rock, Dublin,	Dublin,	22·66
Coniston,	Lancashire,	77·57	Markree,	Sligo,	37·79
Well Head, Halifax,	Yorkshire,	31·57	GENERAL ABSTRACT.		
Leeds,	Yorkshire,	21·69			
York,	Yorkshire,	22·58	England, (mean of all Stations),		35·71
Bishopwearmouth,	Durham,	18·27	Scotland, Do. do.,		38·05
Seathwaite,	Cumberland,	138·46 (!)	Ireland, Do. do.,		35·91
Keswick,	Cumberland,	57·99	The United Kingdom, Do. do.,		36·56

METEOROLOGICAL TABLES. 93

Temperature of some Noted Places.

	Latitude.	Winter's Mean.	Summer's Mean.	Annual Mean.
Europe.				
St Petersburg,	N.L. 59° 56'	18°	61°	39°
Copenhagen,	... 55° 41'	31°	62°	46°
Edinburgh,	... 55° 57'	38°	57°	47°
Berlin,	... 52° 31'	31°	64°	48°
Dublin,	... 53° 23'	40°	59°	49°
London,	... 51° 30'	39°	60°	50°
Paris,	... 48° 50'	38°	64°	51°
Vienna,	... 48° 12'	32°	69°	51°
Constantinople,	... 41° 0'	41°	71°	56°
Gibraltar,	... 36° 7'	56°	73°	64°
Asia.				
Pekin,	N.L. 39° 54'	28°	75°	53°
Canton,	... 23° 7'	54°	82°	69°
Bagdad,	... 33° 19'	49°	93°	73°
Bombay,	... 18° 56'	77°	83°	81°
Calcutta,	... 22° 33'	72°	86°	82°
Africa.				
Cape of Good Hope,	S.L. 34° 11'	58°	74°	66°
Cairo,	N.L. 30° 2'	58°	85°	72°
America.				
Melville Island,	N.L. 74° 47'	−28°	37°	1° (!)
Quebec,	... 46° 49'	+14°	68°	41°
New York,	... 40° 42'	30°	71°	51°
New Orleans,	... 29° 57'	55°	82°	69°
Rio Janeiro,	S.L. 22° 54'	68°	79°	73°
Australia (Hobart Town),	S.L. 42° 53'	42°	63°	52°

INDEX.

	PAGE
Actinism,	74
Aerolites,	64
Agency, Divine,	79
Agents, imponderable,	70
Alcoholic thermometer,—its defects,	21
"*Apparent*" reading of Barometer,	7
Aqueous vapour, invisible,	32
—— its beneficial effects,	57
Aristotle's idea of Heat,	70
Armstrong (Sir William's) Theory,	67
Atmosphere,	1
—— its height and weight,	2
—— its component parts,	60
—— its colour,	72
Atomic Theory,	69
Aurora Borealis,	54
Balloon excursions,	44
Barometer,	3
—— "corrected and reduced,"	7
—— allowance for temperature,	8
—— allowance for elevation,	6
—— dial-faced, and standard,	10
—— diameter of tube,	11
—— hourly fluctuations of,	15
—— how affected by temperature,	9
—— how and where to hang it,	12
—— its indications,	14
—— manner of observing,	14
—— testing before hanging,	11

INDEX.

	PAGE
British rain-fall, Mr Symon's,	47
—— General Table of,	92
Calendar of nature,	77
Callander, rain-fall at,	50
Calm at Equinox,	53
Casella's thermometers,	29
—— new mercurial minimum,	30
Climate, and organic development,	68
Clouds, classes of,	51
Colours, the primary,	74
Comparative Dryness (Summer and Winter),	43
Crieff, elevation of,	6
Cylinders for snow measurement,	48
Daily hour of observation,	24
Daniell's Hygrometer,	34
Decimal Notation,	26
Dew-point Hygrometer,	32
—— by calculation,	39
—— by observation,	38
Dynamical theory of Heat,	70
Earth's rotation, affected by Heat,	18
—— internal heat,	66
—— magnetic sympathy with the Sun,	58
Electricity, atmospheric,	61
English Temperature. (See Table III.)	83
Equinoxial Gales,	53
Etherial Hygrometer,	34
Expansion of water, Law of,	56
Forces, imponderable,	70
Funnel of Rain-gauge,	46
Fusion (subterranean),	67
Gulf Stream,	57
Hurricanes,	54
Hydrogen, presence in the air,	61

INDEX.

	PAGE
Hygrometer, meaning of,	32
—— its variations,	36
—— use in sick-rooms,	45
—— use in hot-houses,	45
—— factors for Dry Bulb,	40
Ice floats, why?	56
—— on the wet bulb,	39
Insects, first appearance to be noted,	77
Law implies Mind,	77
Leslie (Sir John's) Theory,	67
Light, its decomposition,	73
—— theories respecting,	71
Makers of Thermometers,	27
Meteoric Stones,	64
Mistral,	54
Moisture, atmospheric,	41
Ozone, what it is,	63
Ozonometer,	63
Primary Colours,	74
Psychrometric Hygrometer,	35
Radiation, solar,	22
—— terrestrial,	23
Rainbow, cause of,	73
Rain-fall, unequally distributed,	50
Rain-gauge,	46
Range of temperature,	19, 43
Record of observations,	25
Saturation, degree of. (See Table IV.)	84, *etc.*
Simoom,	54
Sirocco,	54
Snow measurement,	48
Standard Barometer,	10
Sun's heat, how renovated,	50

E

INDEX.

	PAGE
Tables. (See end of the volume.)	
No. I. Strathearn Winds,	81
No. II. Six years of Barometer,	82
No. III. Seven years of Thermometer,	83
No. IV. Dew-point and Saturation,	84, etc.
Temperature,	16
—— of the ground,	16
—— of deep mines and springs,	67
—— of the Arctic Regions,	20
—— of noted places, Table of,	93
Thermometers, description of,	27
—— mercury better than alcohol,	21
—— where to place them,	23
—— manner of observing them,	25
Tornado,	54
Twilight, cause of,	73
Tyndall's (Professor) theory,	56
Typhoon,	54
Use of Meteorology,	76
Velocity of Light,	53
—— of Sound,	54
—— of Wind,	54
Vernier, its form and use,	12
Winds, how they modify heat,	17
—— in Strathearn. (See Table I.)	81
Zero, above and below,	25

www.ingramcontent.com/pod-product-compliance
Lightning Source LLC
Chambersburg PA
CBHW020153170426
43199CB00010B/1022